27. 08.

GW 2683939 3

Nineteenth-Century Torpedoes

and Their Inventors

DERBYNIWYD/ RECEIVED	2 8 JUL 2004
CONWY	
GWYNEDD	
MÔN	
COD POST/POST CODE	LL55 1AS

Nineteenth-Century Torpedoes and Their Inventors

EDWYN GRAY

Naval Institute Press

ANNAPOLIS, MARYLAND

GW 2683939 3

Naval Institute Press
291 Wood Road
Annapolis, MD 21402

© 2004 by Edwyn Gray

All rights reserved. No part of this book may be reproduced or utilized in any form or by any means, electronic or mechanical, including photo-copying and recording, or by any information storage and retrieval system, without permission in writing from the publisher.

Library of Congress Cataloging-in-Publication Data

Gray, Edwyn.

Nineteenth-century torpedoes and their inventors / Edwyn Gray.

p. cm.

Includes bibliographical references and index.

ISBN 1-59114-341-1 (hardcover : alk. paper)

1. Torpedoes–History–19th century. 2. Marine engineers–History–19th century. 3. Inventors–History–19th century. I. Title.

V850.G68 2003

623.4'517–dc22

2003017843

Printed in the United States of America on acid-free paper ∞

11 10 09 08 07 06 05 04   9 8 7 6 5 4 3 2

First printing

Contents

Preface and Acknowledgments   vii

Introduction   1

1 Defective in Principle   11

2 The St. Petersburg Connection   26

3 Extreme Simplicity   41

4 A Highly Respected Scottish Engineer   62

5 Most Splendid Results   85

6 Filled with Cork and Glue   106

7 A Perfect Nightmare   129

8 The Wizard of Oz   152

9 This Risk of Premature Explosion   175

10 An Eminently Humane Weapon   194

Appendix | British and U.S. Patents   215

Notes   219

Bibliography   237

Index   241

Preface and Acknowledgments

DESPITE ITS INFLUENCE on battle tactics and its important position in the league table of naval armaments, the torpedo remains one of the world's most underresearched weapons. And its inventors the most neglected. This book will, I hope, go some way in placing the torpedo into its historical context and, in doing so, honoring the many men who contributed to its evolution.

The quest to fill in the gaps that existed in our knowledge of early torpedoes began with my original edition of *The Devil's Device*, written in 1974. The ensuing second edition, published by the Naval Institute Press in 1991, served to expand and enlarge understanding of the weapon's development, and the fact that it was the first *book* devoted solely to the torpedo since Sleeman's *Torpedoes and Torpedo Warfare* of 1880 is mute testimony to the inexplicable disinterest of naval historians in a weapon that has turned the tide of combat on numerous occasions. Indeed, the U-boat campaign of World War I, which was fought mainly with torpedoes launched from submarines, almost determined the outcome of the conflict in Germany's favor. In April 1917, Adm. Sir John Jellico confided to his American colleague, Rear Adm. William Sims, that Britain's food stocks would only last for six weeks if merchant shipping losses continued at the current rate.

The Devil's Device was concerned mainly with the role played by Robert Whitehead in producing and perfecting the "fish" torpedo. But as will be discovered in these pages, many other men contributed to the ultimate success of the weapon. This is their story, and in the course of it, we will follow the story of the torpedo from Robert Fulton's pioneering work in the early 1800s through to 1900, when the weapon's growing complexity renders it an unsuitable subject for a work of this nature which seeks to avoid textbook technicalities only appreciated by specialists.

[vii]

By the turn of the century, many of these inventors had departed the scene, which was by that time dominated by a handful of major armament manufacturers and state-owned arsenals. As a result, development passed from the thrusting single-minded individuals who had stamped their characters on the weapon to anonymous teams of dedicated but faceless engineers—a transition that removed romance from the torpedo story and made my aim to entertain as well as inform an impossible task if I crossed the watershed into the twentieth century. The year 1900 thus seemed a suitable point to bring my narrative to a close.

Some of the weapons appearing on these pages, such as those built by John Lay, John Ericsson, John Howell, and Louis Brennan, will be familiar names to enthusiasts, although only the latter has been the subject of extensive research during recent years—and this mainly in Brennan's native Australia. However, with this honorable exception, an intensive search of archive material, patents, and long-vanished publications has exposed many past misconceptions and errors which are now corrected. Apart from these slightly more familiar weapons, there remained substantial voids in our knowledge of the torpedo's history, especially the numerous devices produced by other now-forgotten inventors which have not seen the light of day for a hundred years or more. Most—and more than seventy examples have been traced—will be completely new to the modern reader. Some of them showed a modicum of promise, others were hopelessly inept, and a few proved to be little more than a cause for hilarity. Nevertheless, successful or unsuccessful, each merits its considered place in the history of naval weaponry.

During research, other matters of interest emerged. For example, despite its acclamation as a brilliant new discovery for enhancing the performance of the torpedo when it was first reported in 1901, the so-called heater principle was, in fact, nearly as old as the Whitehead weapon itself. And inventor after inventor had recognized the advantages of heating compressed air or gas and had come up with practical systems to achieve that object—the first appearing as far back as 1872. Why it took the major torpedo manufacturers so long to "discover" what was clearly a well-understood phenomena remains one of the weapon's unsolved mysteries.

A rocket-propelled torpedo predating the American Civil War was another surprise to emerge from my research. So, too, almost unbelieva-

bly, was a torpedo that had no military use and carried no explosive warhead yet had a practical purpose. Further delving into the archives revealed the name of the man responsible for introducing twin contra-rotating propellers to the Woolwich and Whitehead torpedoes. Indeed, far from being an employee of Woolwich arsenal as stated by Sleeman, a legend that has persisted to this day, he was an eminent and respected Victorian engineer.

Perhaps one of the most interesting items to emerge related to the depth-control gear of the Brennan torpedo, a closely guarded secret for more than 120 years. The chance discovery of an overlooked scientific paper by the Australian academic Prof. William Kernot finally revealed the nature of the mechanism inside the sealed unit containing the depth-control device. This, and the rest of the Brennan story, will be found in chapter 8.

Credit for unearthing many of the facts on which this book is based must go to my Australian colleague, Denis Cahill, from Melbourne, Victoria. Despite living on opposite sides of the world, we have worked together as a team for nearly ten years. Without Denis's dedicated expertise at locating long-forgotten material, this book could have never been written.

Research has been spasmodic rather than continuous, as I have produced a number of other books since the first edition of *The Devil's Device* and over this long gestative period many people have helped with information. Every name listed in both the first and second editions of that book deserves to be repeated again in this note, for all contributed to the sum of my knowledge of the torpedo and those associated with its development.

However, I will restrict my thanks to those who have assisted me in more recent years, after this new book began to take shape: Alec Beanse of the Palmerston Forts Society; Robert Boyle; the late Lt. Cdr. Gervis Frere-Cook, former director of the Royal Navy Submarine Museum; Cdr. John Guard, OBE; the late Count Balthazar Hoyos; Mrs. Dorice Janzer; Geoff Kirby; Michael Kitson; the late Arthur Manns, former curator of the Royal Navy Armaments Museum, Priddy's Hard, Hampshire; Christopher Marks; Dr. Robert Millar, Victoria, Australia; the director and staff of the Royal Engineers Museum, Chatham, and editor of the *Royal Engineers Journal;* David Richards of the Palmerston Forts Society; Cdr. Jeff Tall,

OBE, director of the Royal Navy Submarine Museum, and the always helpful curator, Margaret Bidmead; Dr. Anatol Taras, Belarus; the late Norman Tomlinson, former borough librarian, Gillingham; the U.S. Navy Torpedo Station, Newport, Rhode Island; Jonathan Ward, editor of *Engineering,* London; Dr. C. M. Woolgar, archivist and head of Special Collections, and Gwen McLay, Hartley Library, University of Southampton.

Copyright material used in the text and illustrations is fully acknowledged in the footnotes and is used with the permission of the owners. In a very few instances, due to publishers or periodicals ceasing to trade, it has not been possible to trace the copyright holders. However, the source of the material has been shown in the relevant footnotes. If there remain any omissions in this respect, I offer my apologies and will be pleased to add the appropriate acknowledgments in future editions.

*Nineteenth-Century Torpedoes
and Their Inventors*

Introduction

U NTIL THE advent of the bomb and the guided missile in the twentieth
century, the preceding three thousand years of naval warfare had
been dominated by only four primary weapons. In chronological order,
these were the ram, the cannon, the mine, and the torpedo. Significantly,
three of the four were intended to inflict damage below the enemy's waterline, where a ship was most vulnerable. And of these, the torpedo
proved to be the most flexible and successful. That it continues to be deployed in the age of nuclear war is testimony to its key role in the task of
gaining and retaining command of the world's oceans.

The ram, invented by the Phoenicians between 1100 and 800 B.C.,
along with the oared bireme, which allowed its tactical employment in
battle, was the decisive weapon of the Greek and Roman periods of naval
history, when the sea conditions and geography of the Mediterranean
were ideally suited to the use of large galleys. Being an integral part of
the ship's structure—in effect, an extension of its longitudinal keel—it ter-
minated many feet ahead of the bow and, more important, a fathom or
more below the waterline, where it posed a lethal threat to the lightly
built warships of the classical period. By its very nature, the ram demanded tactical precision for its success, and the controllability of an
oar-propelled vessel which could be steered independently of wind and
tide provided an ideal delivery platform for its effective use in combat.

For this reason it was not a suitable weapon for sailing ships and, consequently, found little favor with seamen operating from harbors along
Europe's Atlantic seaboard. Nevertheless, ramming remained a tactic of
last resort throughout three millennia, and even during the world wars
of the twentieth century it featured in a number of desperate encounters
between destroyers and submarines.

[1]

Ironically, the modern torpedo began life as little more than an explosive version of the ancient ram in the wake of the latter's brief mid-nineteenth-century renaissance, when the arrival of steam power once again allowed men-of-war to be maneuvered independently of the wind. In 1866, at the battle of Lissa, Austria's Rear Adm. Wilhelm von Tegethoff won a famous victory over his more powerful, if less tenacious, Italian enemy by his tactical deployment of the ram to demoralize his opponent. It was, however, to be the weapon's swansong, for only five months after von Tegethoff's resounding defeat of Admiral Carlo di Persano's fleet in the Adriatic, Robert Whitehead's prototype of self-propelled torpedo, an altogether more sophisticated weapon than the explosive ram, was tested by the Austrian navy. Following its adoption two years later in 1868, the mere threat of torpedo attack was sufficient to keep fleets at arm's length, thus denying admirals the opportunity to exploit the ram as a weapon.

Sea fights from classical times up to and beyond the Middle Ages were no more than land battles on water. The objective was to grapple the enemy vessel and board it, the ensuing hand-to-hand struggle determining the engagement in which seamanship played little or no part. Sailors merely handled the vessels in the opening stages of the action, but, once alongside, soldiers were employed for the actual fighting. The long bow, and occasionally large catapults capable of hurling heavy stones or combustible materials, were early predecessors of the gun and were used as stand-off weapons before the combatants closed for the final and decisive stage of battle, although they played little part in determining the outcome of the melee.

The de facto admiral who controlled the battle was usually a distinguished general well versed and experienced in the arts of ground warfare. By contrast, the master of the fleet, the senior seagoing officer, was merely a subordinate whose specialist skills and seamanship were only required to place the ships where the former ordered.

The introduction of gunpowder quickly led to the primitive artillery of the fourteenth century. Although a French cog captured at the battle of Sluys in 1339 is reputed to have been armed with three iron guns, the first cannon to be deployed on an English ship, *Christopher of the Tower*, was not mounted until 1406. These revolutionary new weapons opened the range between opposing vessels in battle, which, in turn, effectively reduced the opportunity of hand-to-hand struggles. Nevertheless, there re-

mained many instances of boarding up to and beyond the Napoleonic Wars, although by the mid-seventeenth century such actions were generally undertaken by seagoing soldiers, or marines as they were termed, rather than rank-and-file ground troops. Command, too, passed from generals to admirals, with notable exceptions. Both Robert Blake and Prince Rupert, for example, were outstanding military leaders before being respectively appointed to command the opposing Parliamentarian and Royalist fleets during the English Civil War (1642–48), though they performed their duties as admirals rather than generals on taking up their naval appointments.

By the middle of the seventeenth century, the gun was supreme and had no competitors as the arbiter of victory in the never-ending struggle for supremacy at sea. It suffered, however, from one inherent disadvantage. Cannon balls and, later, explosive shells, might batter a ship's upperworks, decimate its crew, and knock out its guns, but save for the chance detonation of an opponent's magazine, they frequently failed to sink the enemy. Even during World War II, the *Bismarck* endured a horrendous bombardment of 14- and 16-inch shells which reduced the once-mighty battleship to little more than a burning and defenseless pyre. But the pride of the Kriegsmarine stubbornly refused to go the bottom, despite several torpedo hits, until her survivors flooded the battleship by opening the seacocks. (It should be added that the Royal Navy disputes this version of events.)

On occasion, fire might engulf and destroy a timber-built man-of-war, yet its blackened hulk often remained afloat for days after the battle, and surrender was a common end to many a sea fight when the defeated enemy vessel obstinately failed to sink. Two developments were to change this situation: the coming of the iron warship and the introduction of the explosive shell, both of which emerged in the early to mid-nineteenth century. Lacking the natural buoyancy of a wooden ship, an ironclad could be sunk by shellfire, although it was often a protracted task. And when designers began adding more and more armor to thwart the increasing destructive power of the shell, a contest developed between guns and armor which continued without pause until the end of World War II. Indeed, it was this apparent invulnerability of the ironclad that hastened the development of underwater weapons in general.

Protective iron plating was first fitted experimentally in 1842, and by

1858, France had commissioned four armored but wooden-hulled frigates. Britain retorted promptly by laying down the 9,137-ton *Warrior* and *Black Prince*–the world's first iron-hulled, armored warships–and the race for naval supremacy had begun, although, in the course of time, Germany was to seize the baton from France as the Royal Navy's leading challenger. Britain's complacent reliance on superiority in numbers was exposed at the battle of Jutland in 1916, which demonstrated that better armor protection and more efficient shells could adequately compensate for numerical inferiority.

The American Civil War witnessed a rapid construction of ironclads of all shapes and sizes, though many of them still relied on timber hulls. The conflict also witnessed renewed interest in the use of underwater weapons as designers began to realize that the easiest way to sink an armored ship was to attack it below the waterline, where the hull was unprotected.

Such devices, or so-called infernal machines, dated back as far as 1585, when the Italian military engineer Frederico Gianibelli devised a small surface vessel, constructed from wood, which he filled with gunpowder and other combustible materials and roofed with six-foot-thick tombstones, on top of which were placed cannon balls, shot, lumps of iron, and large marble chippings. A flotilla of these lethal explosive boats were drifted downstream against a defensive barrier the Spanish duke of Parma had erected across the River Scheldt, and, more by luck than judgment, succeeded in blowing a two-hundred-foot hole in the barricade and killing eight hundred of the duke's soldiers in the bargain. Gianibelli's device, however, was certainly not a torpedo. Neither was it a submerged weapon. But it served to direct the thoughts of others in the right direction.

The Dutchman Cornelius van Drebbel produced the first spar torpedo, or water petard, some time after 1618. He is, of course, more famous for constructing the first authenticated submarine, which, according to contemporary reports, was demonstrated before King James I on a stretch of the Thames between Westminster and Greenwich in 1620. It was, however, little more than a semisubmersible, and it ran awash and was probably carried downstream by the current rather than by the exertions of the four rowers inside. Van Drebbel shrewdly realized that the boat would have a greater commercial value if it could be employed for mili-

Introduction · 5

tary purposes, and to this end, he devised a form of spar torpedo which would enable a small flotilla of similar craft to carry out a surprise attack on enemy vessels while they lay at anchor.

Drawings from the period reveal the water petard, as van Drebbel preferred to call it, to have been merely a canister of explosives lashed to the end of a long wooden pole. So far as can be established, and despite the Dutchman's grandiloquent claims, the weapon met with no success during two English expeditions against La Rochelle in 1626, and van Drebbel quickly fell out of favor. He ended his adventurous career as a humble and penurious London tavern keeper, a fall from grace that was to set the pattern for several other notable torpedo inventors of the nineteenth century whose much-acclaimed devices were tested and found wanting.

Underwater weapons played only a minor role in the Napoleonic Wars, but this prolonged period of conflict produced a few new devices as well as the exhumation and improvement of some older ones. The contact mine and the towed torpedo, for example, made their debut at this time, though for some reason the latter invention was not pursued despite sinking a target vessel in 1805 on the one and only occasion its capabilities were publicly demonstrated. Claims that a self-propelled torpedo was tested on the Seine sometime around 1797 are now discounted by historians. A primitive form of limpet mine was devised by Robert Fulton during the Napoleonic period, but the origin of this particular device can be dated to David Bushnell's *Turtle,* which, on 6 September 1776, had unsuccessfully attacked the British frigate *Eagle* while the latter was anchored off the Battery on Manhattan Island. Bushnell had also experimented with the drifting mine, which was to make a reappearance in the opening years of the nineteenth century. However, neither weapon was a torpedo in the modern sense of the word.

It should be noted at this point that "torpedo" was a generic term for all forms of underwater explosives until approximately 1870. From then onward, static devices were identified as *mines* and movable and self-propelled weapons came to be classified as *torpedoes.* There were, inevitably, exceptions to the rule. The towed torpedo, for example, was really no more than a static mine which was dragged to the target at the end of a line attached to a moving vessel. But in view of its short-lived career, the ambiguity can, perhaps, be forgiven.

Underwater explosive "machines" were exploited by the Russians

during the Crimean War (1854–56) against British warships operating in the Gulf of Finland. Designed by Professor Jacobi of St. Petersburg, they were insufficiently powerful to inflict serious damage but are of interest for their use of a chemical detonator—a glass vial of sulphuric acid resting on a bed of potassium chlorate which ignited when broken and caused the main gunpowder charge to explode. The principle was developed later in the nineteenth century to produce the Herz horn, or contact trigger, in which the heat generating fuse produced a small electrical current which activated the detonator.

The development of torpedoes and mines was given a fresh impetus when the American Civil War erupted in 1861. In a conflict between two naval powers, the weaker often has to rely on stealth and subterfuge to overcome its opponent's greater strength. It was thus the Confederate States that sought to employ underwater weapons in its attempt to break the Union blockade of their Atlantic harbors—and to prevent the enemy's riverine incursions. The potential of such devices had already been signaled in 1842, when Col. Samuel Colt demonstrated the use of electrical current charge in New York Harbor. The following year, on 13 April 1843, he successfully blew up a five-hundred-ton brig while the vessel was proceeding down the Potomac River at a speed of five knots. Even more alarmingly, the detonator had been triggered by an electric cable operated from a point five miles away from where the device exploded. In its later and improved form, this type of static weapon was known as an observation mine.

Electrically detonated devices together with other forms of underwater explosive charges, many dating back in origin to the Napoleonic Wars, formed the nucleus of the Confederate navy's armory of submerged weapons. None, however, were self-propelled, although the spar torpedo was at least attached to a small steam-powered launch and driven directly against an enemy hull. During the course of the conflict, the spar weapon and various forms of the static mine employed improved methods of triggering the detonator and the growing sophistication of the technology involved signposted the way to the future. The ultimate breakthrough, the Whitehead torpedo, was only a year away when the North-versus-South struggle finally ended at Appomattox.

On a very different level, the fire ship—an unmanned vessel filled with combustible materials that was floated or sailed downwind against the

opposing fleet–always remained an option if conditions were suitable. Like the ram, it reached back to the Greek and Roman wars. Although a surface weapon powered by wind or current, it was the true predecessor of the modern torpedo insofar as its intent and tactical employment were concerned. With supreme irony it was also, literally, the inspiration for the embryonic Whitehead weapon, which finally emerged in 1866.

At the beginning of that decade, an Austrian marine artillery officer had drawn up the plans of a small surface boat which, like the fire ship, could be loaded with explosives and dispatched against blockading warships steaming close inshore. His name is unknown and his device was never built. On his death, the plans came into the possession of a retired officer of the Austrian navy, Fregattenkapitän (Commander) Giovanni de Luppis, and, with time on his hands, he tinkered with the idea, devised his own modifications, and finally constructed a scale model for testing and demonstration.

Dubbed Der Küstenbrander, or coastal fireship, it was timber-built and propelled by a clockwork motor which turned a screw at the stern. Also at the stern, behind the propeller, was a large rudder controlled by an operator ashore with the aid of rope tiller lines. The gunpowder charge carried inside the hull was detonated by a percussion contact device in the bows. Satisfied that he had produced a war-winning weapon, de Luppis promptly presented himself and his cherished brainchild to the naval authorities in Vienna, who, for obvious reasons, were less than impressed with his rather crude toy. The propulsion system was clearly inadequate for combat operations, and the steering arrangements were considered "unworkable." Nevertheless, the admirals recognized the weapon's potential and advised de Luppis to seek the help of a professional engineer.

A local businessman in Fiume, Giovanni de Ciotta, introduced the retired officer to the technical manager of the town's largest factory, Stabilmento Tecnico Fiumano, an expatriate Englishman by the name of Robert Whitehead.[1]

Although he quickly discounted the primitive contrivance which de Luppis brought to his office Whitehead, like the admirals, sensed instinctively that the coastal fireship contained the germ of a workable idea. However, after many months of hard work modifying the prototype, he realized that the device, as envisaged by de Luppis, was so flawed in concept that it was totally useless.

But by now Whitehead, too, had been bitten by the torpedo bug, and, locked away in a small workshop in the yard of the STF factory, he wrestled with the problem for more than two years before emerging with a revolutionary new weapon that not only ran submerged to the target but also was powered by an inboard pneumatic engine of his own design fueled by compressed air contained within the body shell of the torpedo by means of a high-pressure chamber. Although it bore no resemblance to Der Küstenbrander in either appearance or operation other than the explosive charge and impact detonator, Whitehead generously entered into a partnership with de Luppis and described the new weapon as the Whitehead-Luppis torpedo. In truth, the Englishman had been responsible for virtually 100 percent of the creative technology and engineering skills involved in its creation and construction.

However, despite its superficial external resemblance to the later, very successful Whitehead 16-inch Standard torpedo, the prototype was bedeviled with faults, not least of which was its inability to maintain a steady depth. In addition, its maximum speed of six knots was unacceptably slow, and its range of 220 yards proved woefully inadequate, though both of these latter disadvantages could be overcome with further development work. But Whitehead was in no position to waste time ironing out such problems. He had just achieved national fame as the designer of the engines that had powered von Tegethoff's flagship *Ferdinand Max* to victory at the battle of Lissa on 20 July 1866, and seizing the opportunity, he pulled a few influential strings and persuaded the Austrian navy to test his revolutionary new weapon.

Built from wrought-iron boiler plates and weighing approximately three hundred pounds, it measured eleven feet seven inches in length from the tip of its sharply pointed bow to the end of its tail, and its cylindrical body, "in the shape of a dolphin," had a maximum diameter of fourteen inches. A pair of antiroll vertical fins ran the length of the body shell, and the needle-like nose encased a pistol impact detonator which, on striking the target, thrust back to explode an eighteen-pound charge of dynamite. Powered by a two-cylinder compound oscillating air engine of Whitehead's own design operating on compressed air stored in an air chamber, also built from boiler plate, capable of containing a pressure of 370 pounds per square inch (psi), the twin-bladed propeller revolved at

100 revolutions per minutes (rpm) while a simple hydrostatic valve arrangement actuated the elevators of the depth-control system.

The first trials were staged in December 1866, but the weapon's erratic depth-keeping failed to impress the admirals, though they shrewdly recognized the torpedo's promise. It was clear, however, that further work was needed, and Whitehead returned, somewhat despondently, to his workshop.

Many months later he resolved the depth-keeping problem with a pendulum and hydrostatic device which was to form the basic ingredient of the Whitehead torpedo's success. It was dubbed the Secret, an apt name under the circumstances, for details of the balance chamber were not disclosed to client governments until purchase or licensing contracts had been exchanged. Following further trials in 1868, Austria purchased non-exclusive rights to the weapon, and in 1871 the British government followed suit, setting up facilities at Woolwich arsenal–the Royal Laboratory–to build its own slightly modified versions of the Fiume weapon under license. Recently discovered factory records reveal that 1,083 torpedoes had been sold to nine of the world's navies by the end of 1879. This figure refers only to the Fiume 14- and 15-inch weapons, which laid the foundations of Whitehead's wealth and reputation. No records have been traced relating to the pre-1875 "Standard" 14- and 16-inch models.

Twin contra-rotating propellers were introduced at an early stage, and by the end of the nineteenth century, Whitehead torpedoes were fitted with gyroscopic steering control. The diameter had, by this time, grown to eighteen inches with maximum speeds of 30.5 knots and a warhead weighing around 170 pounds. The maximum range of eight hundred yards remained disappointing, but improved engine technology pushed this up to seven thousand and even twelve thousand yards just prior to World War I.

Whitehead quickly realized that, having solved the depth-keeping problem, his goose was about to lay a clutch of golden eggs, and, astutely, he took the precaution of varying the terms of his original contract with de Luppis. As a result, though the Austrian would continue to receive a share of the income from the prototype torpedo, Whitehead would be free to develop it independently and to retain all of the ensuing profits. It is said that de Luppis died a disappointed man some while later in 1875. But

in fairness, it must be accepted that the concept of the modern torpedo sprang from Whitehead's fertile brain and owed nothing to Der Küsten-brander.

The success of the self-propelled torpedo was worldwide, and it remained a virtual monopoly of the Whitehead company up to and beyond World War I. The challenge to this monopoly and the ways in which inventors sought to circumvent the Englishman's stranglehold over the international torpedo market forms the subject of this book.

CHAPTER 1

Defective in Principle

ROBERT WHITEHEAD'S name was synonymous with the torpedo from 1870 to 1914, and even today, in the early years of the twenty-first century, it is perpetuated by the Italian company Whitehead Alenia Sistemi Subaquei, which produces, along with its other models, the highly successful wire-guided Whitehead A-184 multirole weapon. Indeed, many Victorian writers used his name as a noun and referred to "the Whitehead" rather than "the torpedo." And everyone who read a newspaper knew what they meant. This book, however, is not about the Whitehead torpedo or its inventor, for both have already been fully examined in *The Devil's Device*.[1] This is the story of its rivals. For with the advent of the self-propelled torpedo, underwater weaponry inspired the imaginations and ambitions of inventors, engineers, entrepreneurs, amateur mechanics, and assorted cranks the world over. And Whitehead's erstwhile competitors were only too aware of the vast rewards that awaited anyone who succeeded. Unfortunately, few did.

Most of these torpedoes, when finally traced, proved to be not only hopelessly impractical but in many cases virtually suicidal. And very few progressed beyond the drawing board. Sadly, some did not even get that far and existed only in the fevered imaginations of their creators and in the optimistic, if nebulously vague, wording of a provisional patent application. But for the enthusiast, they provide a rich cornucopia of weird and wonderful weapons, a large proportion of which have not appeared in print for more than a hundred years. Others, while perhaps better known, deserve closer study so that previous misconceptions and factual errors can be corrected and their true story told.

To provide a balanced picture of nineteenth-century underwater weapons, this account has been extended to include both spar torpedoes

[11]

and towed torpedoes. In fact, it encompasses *any* moving weapon intended to sink an enemy vessel by striking it beneath the waterline, where it is most vulnerable. By definition, therefore, it excludes static moored and seabed mines, which until the 1870s were confusingly also known as "torpedoes." It also excludes drifting mines, which only moved subject to the vagaries of the wind, current, or tide and did not approach their target by means of artificial propulsion, together with electrically detonated defensive mines laid to protect harbors and exploded by remote control.

One of the earliest self-propelled underwater weapons, designed by Warsop and Brentnall of Nottingham, England, made its appearance in June 1862, some two years before the Fiume businessman, Giovanni de Ciotta, introduced Robert Whitehead to Capt. Giovanni de Luppis[2] in the hope that the expatriate English engineer could develop the latter's surface-running *Coastal Fireship* into a viable and financially profitable weapon of naval warfare.

Warsop and Brentnall's device apparently ran submerged, but like the later Nordenfelt electric torpedo and others of similar design, it was suspended from floats. Power was provided by a compressed air engine which worked a "horizontal paddlewheel," and, like de Luppis's device, it was steered from the shore by means of tiller lines. An exhaustive search of contemporary technical journals has failed to reveal any further details. No drawings have come to light, and there is no mention of it in the British Patent Office's archives.

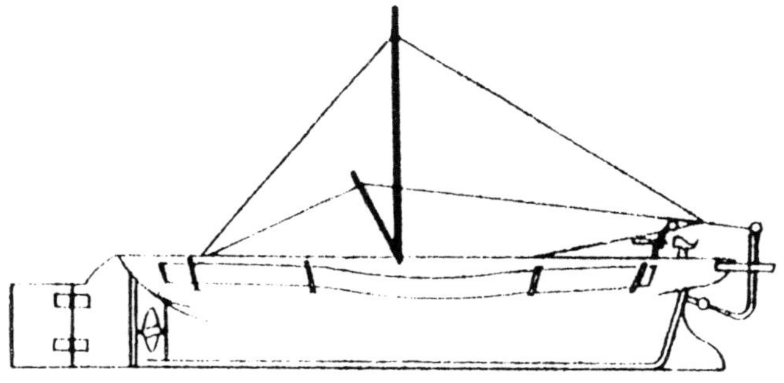

Giovanni de Luppis's Der Küstenbrander, the inspiration for Robert Whitehead's torpedo, as depicted in Sleeman's *Torpedoes and Torpedo Warfare* (2d ed.).

Defective in Principle · 13

It is clear, however, that a working model was constructed, for a proto-type was tested at Portsmouth by Capt. R. S. Hewlett and officers from the Royal Navy's gunnery establishment, HMS *Excellent,* in June 1862. According to the report of the Admiralty's Torpedo Committee, published in July 1876, it "failed to achieve its purpose and was considered so defective in principle . . . as not to merit further trial."[3] This view was, not surprisingly, disputed by one of the torpedo's inventors, George Warsop, who on learning of Whitehead's contract with the British government, was moved to make a public protest to the influential journal the *Engineer:*

I beg to inform you that in 1862 I constructed and tried at Portsmouth a torpedo on precisely the same principle as Mr Whitehead's. [This was not strictly true, but the press reports were so vague that Warsop was misled into believing that the Whitehead torpedo was identical to his 1862 test model.] . . . The after portion . . . served as a reservoir for highly compressed air, the initial pressure of which was about 600 psi. [The 1866 Whitehead had operated at a pressure of 370 psi, and it was noted that American engineers were unable to harness this pressure four years later when they tried to copy the weapon. If Warsop's claim was correct, it meant that some aspects of his torpedo were extremely advanced.] This served as the motive power for a small engine actuating a simple feathering paddle on the outside of the torpedo . . . [which] could be propelled at any depth, at great speed, and its course indicated by a small float on the surface of the water, was always under the control of the assailant [and could be] unerringly steered to its mark.[4]

From this explanation, it is apparent that Warsop relied on floats and had devised no mechanism for controlling the running depth of this weapon, a problem Whitehead struggled to resolve for several years prior to the second, and successful, Austrian trial of 1868,[5] while the "feathering paddle-wheel" seems unlikely to have been capable of producing the "great speed" claimed by its inventor. With perverse logic, a fault from which several other torpedo inventors suffered, the use of the float on the surface, often surmounted by a small flag or disc, negated the torpedo's prime advantage of running submerged and invisible when attacking its target.

No doubt influenced by the battle of Lissa, fought in 1866, Victorian tacticians thought of naval actions in terms of a melee rather than a gun

battle between disciplined lines of warships. And so the idea of an uncontrolled free-running submerged weapon missing its intended target and sinking a friendly vessel in the hurly-burly of combat caused considerable alarm to both admirals and armchair pundits. Indeed, the *Engineer*'s editorial on the British government's purchase of the Whitehead–de Luppis weapon was very antagonistic:

> It is of the first necessity that weapons of war should be harmless to those who use them, for the dread of being "hoist by their own petard" is apt to unnerve the best of men, and thus bring about the very evil anticipated. Serviceable torpedoes can be made as perfectly innocuous to those who employ them as shot and shell used with great guns; and none other should be introduced unless for very exceptional cases, in which friendly vessels cannot be implicated.... If [the] £15,000 given or guaranteed by the Government for this uncontrollable [Whitehead] weapon had been spent in experimenting upon and perfecting the two naval torpedoes now in service the money would have been more usefully employed. These two are the perfectly safe outrigger [or spar] torpedo . . . and the mechanically ignited Harvey towing torpedo. [It] is of the nature of the fish torpedo, however perfect its action, to be highly dangerous and indiscriminating, and as liable to destroy its friends as its foes.[6]

In his later years, Admiral Lord Fisher was fond of quoting a former first sea lord, thought to have been Adm. Sir Sydney Dacres, who, he claimed, had once told him that "there were no torpedoes when he came to sea and he didn't see why the devil there should be any of the beastly things now."[7] It would seem that this view, perhaps more elegantly expressed, was shared not only by Sir Sydney's fellow admirals but also by the engineering profession and the technical press.

Despite his boast that he had almost singlehandedly introduced it into the Royal Navy, Fisher in fact was similarly opposed to the fish torpedo in its early days, despite knowing more about its abilities and performance than most of his superior officers, for he had first become acquainted with the Whitehead weapon when he was given private access to confidential technical reports by Capt. Louis Hassenpflug while on an official visit to Kiel in August 1869. The latter's inside knowledge is unquestionable. The son of the Grand Duchy of Hesse's prime minister, Hassenpflug

Defective in Principle · 15

later became Robert Whitehead's son-in-law when he married the inventor's eldest daughter, Frances Eleanor.

Enclosing copies of his notes, Fisher reported to the Admiralty on the twenty-first of that same month, "Once started [Whitehead's] torpedo is beyond control and will blow up any one it comes across."[8] In the same report, he claimed that the Harvey towed torpedo was "a far superior" weapon, a view he clung to for the next three years while he was serving abroad in Far Eastern and Australian waters on HMS *Ocean*. It is curious that Fisher, a gunnery specialist, did not equate the uncontrolled track of a torpedo with the equally free trajectory of a naval shell. The fact that the torpedo, like the gun, was aimed seemed to have eluded him.

Faced by such an antagonistic climate of opinion, it is not surprising that Whitehead's competitors sought to overcome the fish torpedo's inability to make a deliberate alteration of course after it had been fired, and most of the designs that followed incorporated, for better or for worse, some form of guidance system. But to keep these later developments in context, one further backward glance is necessary.

In the same edition of the *Engineer* that featured George Warsop's letter, there was another, this time from Philip Braham of Bath, who wrote to say that he had submitted "a scheme" to the Admiralty on 2 July 1868 without success. Reference to the drawings published in the same journal on 20 October 1871 following the reading of a paper by Braham to the British Association reveals his "torpedo" as totally impractical, being no more than a passive explosive charge expelled from a submerged launching tube by compressed air and propelled through the water by the impetus of the initial discharge. And it is clear that Braham either intended to use it at suicidally close range or, more likely, had given little thought to the resistance of the water through which it was to travel after launching. The projectile, like that of John Ericsson's many years later, was made from wood with an "explosive chamber" at the front within a corrugated iron cap, the latter being required to balance the weapon in a horizontal plane. At best it was no more than a detachable explosive ram to be employed at arm's length from its target and was probably useless beyond a range of twenty yards or so.

The Admiralty politely informed the inventor that "although they consider [his] method shows considerable ingenuity, they are not of the

16 · *Nineteenth-Century Torpedoes and Their Inventors*

opinion that it is likely to be of use to Her Majesty's Navy."[9] And when Braham pursued his request for a trial, he received a rather more tart response from Their Lordships who were "not desirous of carrying out experiments in the manner you propose."[10]

A third letter to appear in the *Engineer* on the same date came from Andrew Alexander of Worcester, who had patented an "exploding submarine missile" on 27 July 1864.[11] To use modern parlance, this incorporated a self-destruct mechanism which would detonate the warhead at the end of its run, thus overcoming the Victorian dread of an armed torpedo running amok and threatening friendly vessels with instant annihilation. But Alexander's invention appeared to be merely theoretical, and he did not proceed beyond his application for a provisional patent. Indeed, he admitted in his letter that he had "spent little cash or time" on his torpedo and readily accepted "that calculations based on [such] uncertain data may be worth very little."[12] At least Andrew Alexander, unlike many other contemporary inventors, had the merit of being honest.

His proposed weapon was one of the first to be propelled by a rocket, the exhaust gases of which would "impart an onward rotary movement." Alexander stated that he preferred this method to his original option of compressed air. It is also of interest as being the earliest "with spiral vanes on the exterior thereof to impart rotary motion thereto." But when he conjectures that it would have a range of half a mile "at the speed of a fast express railway train," the weaknesses of his theoretical extrapolations become embarrassingly apparent.[13] Whitehead's 1866 prototype, tested two years after Alexander's provisional patent, boasted a speed of just six knots and a range of 220 yards. Harsh reality was often a long way behind an inventor's daydreams.

The Admiralty quickly passed the papers to the War Office, where, seven years later, Alexander lamented, "It possibly still lies hopelessly blocked up in one of the remote sidings of that bewildering department."[14]

George Quick of Portsea was even more optimistic than Alexander, for his "self-propelling torpedo" was said to be capable of covering two thousand yards in a mere thirty seconds and was described as having a diameter of twenty-four inches with a warhead containing seven hundred pounds of gunpowder. As an added attraction, it could run submerged in "all conditions of weather" at depths of up to twenty feet. The War Office

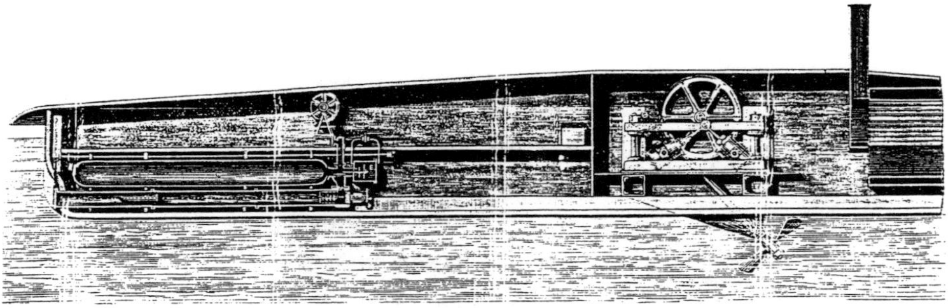

Philip Braham's unpowered inertial torpedo of 1871 pointed the way to John Ericsson's later projectile weapon and his torpedo boat *Destroyer.* Neither were successful.

tested the 135-mph torpedo at Chatham, but as nothing further was heard of either Quick or his weapon, it must be presumed that it did not live up to his expectations.[15]

Another form of projectile torpedo was produced in the United States by a former U.S. Navy officer, Robert Weir, in 1870. Very few details of this particular device are known, but a report in *Engineering* confirms it to have been rocket-propelled.[16] Unlike most of its contemporaries, however, a wooden scale model was not only built but also tested. Its performance was less than impressive. Running submerged at a depth of two feet, it struck a target which had been conveniently placed a mere twenty yards from the firing point. However one looked at the achievement, it was scarcely an earth-shattering event. Not surprisingly, nothing further was heard of Weir's underwater projectile.

With the exception of the Warsop and Brentnall device, none of the weapons examined offered the comforting safeguard of the guidance system deemed so important by naval analysts, and it was not until Ericsson's pneumatic torpedo arrived on the scene that this particular omission was rectified. Whitehead, of course, was unconcerned by the furor in the English press, which continued to claim that his torpedo was as great a threat to its friends as to its enemies. He had shrewdly realized at an early stage of development that a torpedo which held its course and was aimed correctly provided its own built-in safety factor, but as a sop to his critics, he quickly devised a means of allowing the weapon to sink to the bottom at the conclusion of its run. For exercise purposes, when the

torpedo was carrying a dummy warhead, the procedure was reversed and the weapon rose to the surface so that it could be recovered and used again.

Most sources date John Ericsson's pneumatic torpedo to 1873, but a report in the *Engineer* referred to "a recent trial" of the weapon in the Bay of New York, which means it must have been completed during 1872 at the very latest.[17] A letter from Ericsson to another journal, however, suggests a much earlier date for the inventor claimed that he had provided details of the weapon to the king of Sweden and Norway in November 1866.[18]

Ericsson was one of the nineteenth century's outstanding inventors, and his contribution to torpedo warfare and the screw propeller will be considered in chapter 4. At this point it is sufficient to note that his "pneumatic torpedo" lived up to its name in several ways and differed from the John Lay and Victor von Scheliha weapons by "the absence of electricity or magnetism."[19] Its double-cylinder oscillating engine was driven by compressed air supplied from an external source via a tubular cable or hose, unlike the Whitehead, which, though employing a similar power unit, stored the air in a pressurized chamber inside the body of the torpedo.[20] By varying the air pressure it was possible to steer the weapon and control its speed.

The ungainly body comprised a rectangular box made from thin steel plates thirty inches deep, twenty inches wide, and eight feet six inches in length. It was launched by means of deck davits as its cumbersome shape made it unsuitable for tube firing like the sleeker Whitehead torpedo. Even a superficial glance at photographs of the two weapons at this stage (1870) of their development makes the comparison that of chalk and cheese. The pneumatic torpedo was in reality a typical Victorian monstrosity, especially when viewed alongside the Whitehead, which in its essentials was to change little over the next one hundred years of development.

Ericsson's pneumatic torpedo proved disappointing in practice as well, and it was soon found "that the drag of the hollow cable carrying the air from the compressor to the torpedo, so interfered with the steering that the [weapon] was nearly unmanageable."[21] Similar in concept to the original de Luppis prototype, though much more technically advanced, the Ericsson pneumatic was a surface-running torpedo. The warhead contained four hundred pounds of nitroglycerine and the entire weapon

Defective in Principle · 19

John Ericsson's pneumatic torpedo, circa 1870. Note twin two-bladed propellers at the stern.

tipped the scales at some two thousand pounds. The report in the *Engineer* of 10 January 1873, following the New York Bay trials, refers to a tubular cable measuring half a mile in length. As this was the means for both supplying the compressed air that fueled the engine and providing the pressure for steering and speed control, it must be assumed that the maximum range of the Ericsson weapon in late 1872 was just under 880 yards. The Fiume standard 16-inch torpedo of 1871 had an extreme range of about a thousand yards, so there was little to choose between them in this respect. But when it came to speed the Ericsson had a distinct advantage, having been tested at ten miles an hour, while the contemporary Whitehead model, as built by the Royal Laboratory, started out at around seven knots maximum, although tinkering by the Woolwich mechanics, notably the foreman, Lowe, had raised this to an average of ten and a quarter knots over two hundred yards by late 1874.[22]

The adoption of the contra-rotating twin propeller system by Woolwich in that same year, however, pushed the Whitehead up to an average

speed of 12.33 knots for two hundred yards, which restored its superiority in performance.[23] By a strange quirk of fate, the contra-rotating propeller system had been invented by Ericsson himself in 1836,[24] and he had fitted it to his own pneumatic torpedo by the beginning of 1870 or, possibly, even earlier. Although no internal plans of the pneumatic torpedo have been traced, the ungainly dimensions of the weapon suggest that Ericsson had not developed the bevel gearing that enabled the Royal Laboratory to squeeze the contra-rotation mechanism into a body shell with a diameter of only sixteen inches. Unraveling the myths that surround the adoption of contra-rotating propellers by the Royal Laboratory produced a fascinating insight into nineteenth-century engineering rivalries and will be told in detail in chapter 4.

The U.S. Naval Torpedo Station, which had been established on Goat Island, Newport, Rhode Island, in 1869, made an early attempt to copy the Whitehead weapon and it was demonstrated before a select band of specialist officers in 1871. The torpedo, when unveiled, had a diameter of fourteen inches and a length of twelve and a half feet and bore a very close resemblance to the Fiume-built Whitehead model in its external appearance. The propulsion system, however, differed considerably and comprised a two-cylinder engine with the axes of each cylinder parallel to the axis of the torpedo—Whitehead by then was using a Vee-twin configuration of his own design—and the linear motion of the two pistons was transposed to a rotary movement by means of a grooved drum which was geared down to the propeller shaft. Although the diving apparatus was said to have worked reasonably well, "generally speaking the results of the trial were unsatisfactory."[25] Subsequent comments suggest that the body shell was not completely watertight and the air chamber proved unable to contain the required pressure.

A second attempt followed a short while later, but this time the new weapon was not tested in open water but remained moored to the quayside, a decision which suggests the U.S. Navy had little faith in it. Learning from the mistakes of these first two prototypes, Lt. Francis Morgan Barber produced plans for an improved fish torpedo which were submitted to the Bureau of Ordnance in June 1874. Although the design was notable for its bronze air vessel, which Barber was forced to adopt because at that time American industrial know-how did not extend to steel

Defective in Principle · 21

chambers of the required strength, the weapon was not taken up by the bureau and, presumably, was written off as yet another failure.

Barber also built a rocket torpedo in 1873. Constructed from a cast-iron tube, wrapped in asbestos, and sheathed with oak, it had a diameter of twelve inches and a length of seven feet. But its performance was poor and unreliable, due in part to its overall weight of 287 pounds, or a fraction over two and a half hundredweight, and like many other similar projectile weapons, it too has been consigned to history and lost without trace.[26]

Another American rocket torpedo was reported two years earlier in the 17 February 1871 issue of *Engineering*. Designed by Mills L. Callender of New York City, it was said to be an improved version of a marine rocket which he had devised during the Civil War and had patented on 16 October 1862.[27] It had a modernistic cigar-shaped body with a combustion chamber at the bow end. The gas released on ignition was vented rearward from the underbelly of the weapon thus impelling it forward. The "torpedo"–or, more aptly, the warhead–was accommodated amidships and angled down at 45 degrees and attached to a stout chain which was, in turn, secured to the nose of the weapon. When the projectile struck the target it ignited a further explosive charge inside the main body of the "torpedo" which thrust forward the explosive charge. As this was still secured to the chain its downward plunge was quickly arrested and it then swung in an arc to detonate against the keel of its victim, the rigidly tight chain acting as a pivot. Although described as an improvement, there appeared to be little difference between the 1862 and 1871 versions. All in all it sounded more than a little impractical and, hardly surprisingly, nothing further was heard of either the device or Callender.

However, the previous year, 1870, witnessed the birth pangs of a torpedo that was in many respects technically equal to the Whitehead and yet owed nothing to the Fiume weapon so far as propulsion and lateral control were concerned. It had been invented by Lt. Cdr. John Adams Howell, USN, and it relied on the hitherto unexploited principle of the flywheel as the source of its propulsive power. The initial design was submitted to the Bureau of Ordnance in June 1870, but that August body evinced lukewarm enthusiasm for the idea. In its wisdom, however, the bureau agreed to let the young officer construct a model for practical testing.

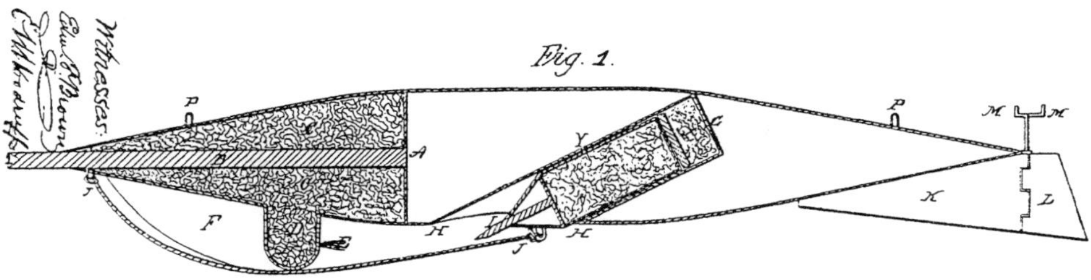

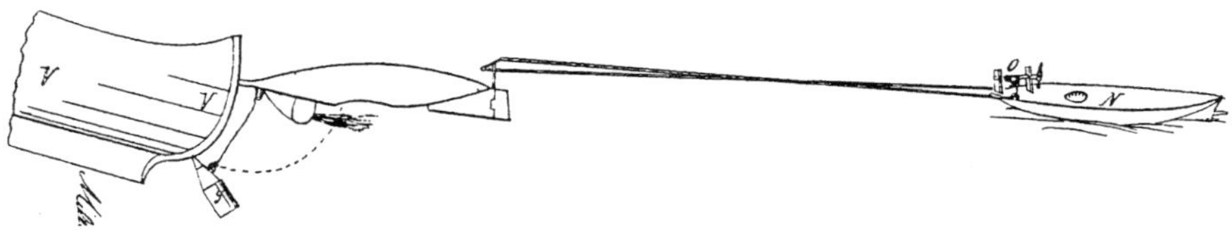

22 · *Nineteenth-Century Torpedoes and Their Inventors*

Fig. 1.

Fig. 2.

Although patented as early as 1863, Miles Callender's separate pivoting explosive charge was copied by several other inventors during the nineteenth century.

Virtually all previous accounts have described the first Howell torpedo as being eight feet in length with a diameter of fourteen inches. This misconception is probably due to the weapon's incredible twenty-year period of development which led to subsequent misunderstandings about the official numbering system adopted by the U.S. Navy. It was 1890 before the Howell achieved the standards demanded by the Bureau of Ordnance and on finally entering service it was designated the Howell Torpedo Mark 1. As a result many writers have assumed that Mark 1 referred to the first Howell to be built rather than the first service pattern. We are at this point only concerned with John Howell's early work. Details of his career and later improvements to the weapon will be found in chapter 8.

By comparison with those later developments, the initial prototype of 1870 was a poor and rather inadequate device–just four feet in length and with a diameter of only twelve inches. Propelled by the energy released by its spinning flywheel, which was activated externally shortly before launching, it had two propellers, one at each end, with both the flywheel and the warhead contained within its cylindrical body shell. Thus, like the Whitehead, it was completely self-contained and, once launched, not subject to external control.[28]

Defective in Principle · 23

Encouraged by the initial test runs, and undeterred by the lackluster responses of the bureau, Howell decided to build a full-sized prototype at his own expense which was completed in 1871. But further trials revealed that the original flywheel configuration, with its axis parallel to the longitudinal axis of the torpedo, was unsatisfactory, and Howell subsequently adopted a new arrangement which positioned the flywheel in a vertical plane with its axis at right angles to the longitudinal axis of the torpedo.

Fortunately for historians, Howell patented this full-sized prototype, and the drawings confirm the horizontal axis of the flywheel.[29] However, in the specification he indicates that he was already considering a different and improved arrangement, for he noted, almost as an afterthought, that "practically, it might be better . . . that the axis of the flywheel should be perpendicular to the torpedo." He also claimed, "My invention is to apply that portion of the inertia of the revolving wheel, which tends to keep its axis of rotation unchanged, to the counteracting of deviating forces," a clear indication that he had already noticed and appreciated the properties of the flywheel's gyroscopic effect.

The accompanying drawing also shows that the propellers at each end, mentioned by Milford, had now been discarded. And while the drawing does not indicate the location and number of propellers employed on the full-sized prototype, the specification refers to the action of the flywheel "acting through a screw," thus confirming that, at this stage of development, Howell had adopted a single propeller.

Commenting on a later and more sophisticated version of this improved layout, the *Engineer* reported that "the Howell torpedo, by its gyroscopic principle, is the only torpedo [to be] entirely automatic in maintaining its direction. It has no ballast, but after launching, automatically takes the depth for which set, and directs itself in a vertical plane. . . . It steers itself automatically, though not in the generally accepted sense of boat steering. Ordinarily a boat is steered on a course by using its rudder to return to such course when forced off it by any agency; but the [Howell] torpedo, when acted on by similar forces, simply rolls to right or left instead of changing course, and this rolling causes the regulator to give a series of impulses to vertical rudders, which produces a resultant motion of the torpedo, opposite to that given by the exterior deflecting force. The result is that the torpedo, having been rolled by a deflecting force, is

rolled back to its normal position by the automatic action of the rudders, there having been no change in the original direction of course."[30]

In his splendid series on torpedo development in *Engineering*, Cdr. Peter Bethell seemed unaware of the details of Howell's 1870 prototype, for his text refers to it as having twin propellers operating side by side and with the axis of the flywheel being set *horizontally* and at right angles to the axis of the torpedo.[31] As already noted, the latter arrangement was not employed in Howell's first design but was adopted subsequently following the unsatisfactory results of the full-sized torpedo's trials in 1870–71. The twin screws did not appear in his patent drawings until 1885, though they may well have been used experimentally some years earlier.[32]

Bethell's account, however, helps to clarify the function of the transverse azimuth pendulum control system which Howell adopted, and his description really cannot be bettered. Having pointed out that a spinning wheel or flywheel "does not so much resist an influence to alter the direction of its axis as respond to that influence . . . [thus] any influence tending to deflect the Howell from its [preset] course would make it heel over. The inventor took advantage of this by installing a pendulum suspended on a fore-and-aft axis. As the pendulum swung to one side or the other, it would apply a correcting helm in the appropriate sense to the vertical rudders . . . [and] the system also acted as an anti-rolling device."[33]

It is remarkable that the two most successful inventors, Robert Whitehead and John Howell, both made use of the pendulum and yet applied the principle for two entirely different purposes. Howell employed it to correct deviations in course, while Whitehead incorporated it into his balance chamber as part of his famous Secret for maintaining the torpedo's running depth.[34] Howell's coincidental use of the gyroscopic effect of the flywheel to maintain a steady course was not matched by Whitehead until 1895, when he adopted the Obry gyroscope after earlier experimentation with the Petrovitch device a few years earlier.[35]

The Howell torpedo's final and unpremeditated advantage was the absence of the telltale line of exhaust bubbles which rose to the surface to reveal the track of an approaching Whitehead-type weapon, a phenomenon that resulted from using compressed air as a propellant. This partic-

ular problem was not overcome until the advent of the electric torpedo in World War II. Further details of the later Howell weapons and the inventor's naval career will be found in chapter 8, for the flywheel torpedo remained in operational use by the U.S. Navy until around 1905, although by then, the American license-built version of the Whitehead was the dominant and unchallengeable underwater weapon.

CHAPTER 2

The St. Petersburg Connection

TORPEDO HISTORIANS are apparently completely unaware of Col. Victor von Scheliha or of his contribution to the weapon's development. This is perhaps hardly surprising, for he has only been independently mentioned in print on four occasions: in a report on a series of lectures delivered by A. L. Alliman to the Kings College Engineering Society in 1877–79 in which his torpedo was incorrectly referred to as "the Sheliba," as a subject in the 1876 report of the Admiralty's Torpedo Committee, in a letter he wrote to the *Times* on 9 October 1874, and finally, in a volume of memoirs by Col. the Hon. Frederick Arthur Wellesley published posthumously in 1932 by his son, Sir Victor Wellesley. However, he had served as a senior military engineer in the Confederate army during the Civil War and had written *A Treatise on Coast Defence* in 1868. Fisher, too, was clearly aware of him and mentioned his name on several occasions in correspondence, while the archives of the British Patents Office held several of his patents for railway signaling devices as well as for the torpedo itself. At this juncture in history, von Scheliha had taken up residence in St. Petersburg following his sojourn in America, but in 1871 he emerged from the shadows to stake his claim as one of the most important early pioneers of the torpedo.

The fact that his weapon was considered equally with those of Whitehead, Ericsson, and Lay by the Admiralty's Torpedo Committee, which met for the first time in May 1873 at the Royal Naval College in Portsmouth, England, is an indication of its contemporary importance. The existence of this committee is rarely mentioned by historians, so a few details of its findings need to be placed on record.[1]

[26]

The committee was chaired by Capt. Morgan Singer, a member of the original team that had appraised the Whitehead torpedo during the trials off Sheerness in September and October 1870 and recommended the acquisition of its manufacturing rights.[2] Sitting with him, among others, was Capt. Henry Boys of the Royal Navy's gunnery establishment HMS *Excellent*. Although not senior enough to be a member of this elite group, Cdr. John Fisher, the newly appointed head of the Torpedo School, was requested to give his expert advice and assistance to the committee when necessary.[3]

The primary purpose of the enquiry was to determine the best means of operating "offensive" torpedoes against the enemy and, conversely, how best to protect the Royal Navy's ships "both at sea and at anchor" from the depredations of the enemy's own underwater weapons. The use of the sea mine and other "floating obstructions" together with ascertaining the most efficient manner of clearing a passage through the torpedo defenses of an enemy also fell within the committee's remit.

Captain Singer's preliminary report was issued in October 1873, though publication of the final version was delayed until July 1876. Having compared the respective performances of the single-propeller torpedo and the improved twin contra-rotating propellered model—to the clear advantage of the latter—the committee noted with satisfaction that "the torpedo could [now] be made to 'sink' or 'swim' at the termination of its run." This new facility removed the main objection to the Whitehead weapon, namely the perceived danger of the torpedo striking a friendly vessel should the weapon miss its intended target. Fisher, of course, was a leading critic of the Whitehead on this particular score at this time, although he was to change his mind within a few months.

The committee also observed that among the "improvements made at Woolwich have been the introduction of a second propeller immediately abaft the first, the screws revolving in opposite directions by means of gearing on concentric shafts, *and the removal of both propellers to a position immediately abaft the horizontal rudder*."[4] This latter modification, though intended to improve performance, makes it easy for even a non-expert to distinguish a Woolwich torpedo from its Fiume parent, as Robert Whitehead steadfastly refused to adopt the Royal Laboratory's tail configuration.

However, the committee's final report seemed more concerned with static mines than with locomotive torpedoes, although it was sufficiently impressed with the self-propelled weapon to warn that torpedoes were "specially inimical to the manoeuvring of large squadrons, and as having a tendency *to reduce to one common level the Naval Power of the greatest and the most insignificant nations.*"[5] It also spelled the end of England's traditional policy of close blockade, for the committee "had no hesitation in expressing their opinion that none of our large vessels could remain for any length during war off an enemy's port without the imminent risk of destruction by offensive torpedoes." It was a conclusion ignored by the majority of the Royal Navy's senior admirals, including Fisher himself, who, as late as 1912, was still agitating for a close blockade of Germany's North Sea ports, and somewhat incredibly, even her Baltic harbors. Fortunately wiser counsels prevailed and the more sustainable policy of distant blockade was adopted shortly before the outbreak of war in August 1914. But let us return to Col. Victor von Scheliha.

The British Patent Office records revealed a provisional patent, dated 25 November 1872, in the name of the Hon. Frederick Arthur Wellesley, of 20 Albermerle Street, in the county of Middlesex, which had been lodged on behalf of Colonel von Scheliha of St. Petersburg, Russia.[6] It seemed probable that Wellesley, on the basis of his name and title, had connections with the duke of Wellington. And further investigation revealed that he was, indeed, a nephew of the Iron Duke, being the third son of the first Earl Cowley, who, in turn, was a brother of Arthur Wellesley, England's greatest general, the victor of Waterloo, ennobled as the duke of Wellington, and prime minister of the United Kingdom from 1828 to 1830.[7]

So how did an obscure German army railway engineer living in Russia come into contact with a member of the illustrious Wellesley family? The answer turned out to be more prosaic than anticipated. Frederick Wellesley, a colonel in the Coldstream Guards, had retired from the army before he reached the age of thirty and moved into the diplomatic service. With his family connections, he was admirably placed for such a career change. Appointed as first secretary of the British embassy in Vienna, he was soon afterward posted to St. Petersburg to serve, appropriately, as the military attaché.

The St. Petersburg Connection · 29

According to Wellesley, "When I first arrived in Russia I made the acquaintance of a retired German engineer officer, a certain Colonel von Scheliha, the inventor of a torpedo steered by electricity, which he desired the Russian government to adopt."[8] The weapon was clearly more than just a drawing-board dream, for Wellesley noted that he "was present at several trials of [von Scheliha's] invention on the Neva [River]. All of which were most satisfactory."[9] As a senior military man it must be assumed that his assessment was objective. It was apparently also a viable proposition in the eyes of the Russian government, which, aware that von Scheliha had no working capital, generously offered to build the weapon in its nearby naval arsenal if he would provide the necessary drawings and specifications. Von Scheliha, a simple soldier, was delighted at his good fortune and saw no danger in accepting an official governmental proposition. And having handed over his working papers to the Russian Admiralty, he followed the advice of the czar's senior officers and "took up his abode in one of the most fashionable hotels in the capital and entertained the heads of departments and their wives lavishly in the hope of obtaining their support" for the project.[10]

An examination of the provisional patent reveals many similarities to the Whitehead weapon; indeed, at first sight it appears to be no more than a somewhat garbled hearsay description of the Fiume-built weapon. But a more careful examination reveals some very advanced scientific and technical thinking. The engine, for example, employed "three brass cylinders placed at a distance of 120° from centre to centre" and was powered by compressed air stored in an inboard pressure chamber.[11] The contemporary Fiume standard 16-inch torpedo utilized a Vee-twin engine of Whitehead's own design and a three-cylinder unit was not fitted until the Fiume 14-inch Model A, production of which began in 1874, the first weapon being delivered to Germany on 26 January 1875.[12] Even then, the three-cylinder engine used in that model was the brainchild of Peter Brotherhood and not of Robert.[13]

In addition, von Scheliha's torpedo was steered by electricity fed through an external cable, the first to employ this source of power for a guidance system, though it is possible that John Lay may have beaten him to it. The latter's contribution to torpedo development is the subject of chapter 3. Ericsson's pneumatic torpedo relied on compressed air

delivered by hose from an external source to steer and otherwise control its course while neither Whitehead nor Howell were in favor of any form of guidance system. Ericsson, too, supplied his pneumatic engine with compressed air through the same hose whereas both the Whitehead (compressed air) and the Howell (flywheel energy) were, like von Scheliha's prototype, completely self-contained.

The torpedo's unique guidance system was as ingenious as one would expect from a senior army railway engineer with a specialist interest in the use of electricity. Alongside his work on torpedoes, he also took out several patents for improvements in signaling circuits and, despite retirement, was obviously a man with a lively and inquiring mind. Even without drawings, which were not included in his application, the text of the provisional specification of his first torpedo is a model of clarity:

> Within the torpedo is a reel on which 4000 or 5000 yards of insulated wire can be wound, and this wire is drawn from the reel as the torpedo advances. When electrical currents are transmitted through this wire, they excite electromagnets within the torpedo, and cause them to attract or repel a permanently magnetic armature. A lever in connection with the armature supports two rods with pawls upon them to act upon two catcher wheels. The rods receive from the engine a to-and-fro movement but so long as the armature is not attracted or repelled by the magnets, the pawls do not come in contact with the teeth of the ratchet wheels. Immediately however that the magnets act upon the armature, one of the pawl rods is lowered while the other is raised, and the former then commences to turn its ratchet wheel, and so drives a worm and rotates an arc with which the rudder is coupled.

The inventor then points out that a reversal of the current similarly reverses the position of the armature which consequently moves the rudder in the opposite direction. This guidance system answered the criticisms of the pundits who objected to the lack of control offered by both the Whitehead and the Howell with the consequential imagined dangers of a torpedo running amok after overshooting its target. The only other contemporary weapon of any competence with a guidance system was Ericsson's pneumatic torpedo, but this was not a self-contained device as its compressed air propellant was supplied by an external source. On the basis of his electrical guidance system alone, von Scheliha's torpedo re-

The St. Petersburg Connection · 31

flected a technical advance on the Whitehead. And, of course, most heavyweight torpedoes in operational service in the first decade of the twenty-first century are wire-guided.

The next section of the patent specification, which is concerned with depth-keeping and an antiroll device, seems to be based on an assortment of ill-digested rumors concerning the details of Whitehead's own depth-control system. The torpedo "is maintained in a horizontal position . . . by means of horizontal rudders or 'fins' which are controlled by suspended weights. . . . In place of controlling the 'fins,' by suspended weights, they may [instead] be controlled by a spring and diaphragm, or other such apparatus which is influenced by the pressure of water to which it is exposed. When this pressure of water upon the apparatus is in excess of a certain amount it inclines the 'fins' upwards, and the torpedo rises similarly." He continues that when the water pressure decreases, the "fins" are inclined downward by the opposite action of the apparatus so that the torpedo sinks deeper into the water.

Bearing in mind that a "suspended weight" is an apt synonym for a pendulum and that the "spring and diaphragm" mechanism equates in broad terms with the hydrostatic valve, von Scheliha appears to be describing in his own words the balance chamber which Whitehead had dubbed the Secret when it was introduced in 1868. Pirated though it may have been, it must be remembered that Wellesley, who had observed the practical tests of the torpedo, confirmed that it worked. And the colonel was certainly not the only person to "borrow" from Whitehead over the years.

The most astounding part of the provisional specification, which certainly owed nothing to Whitehead, came at the end, where von Scheliha added, almost as an afterthought, "In some cases I arrange a small furnace within the compressed air reservoir, and ignite the fuel at the time that the torpedo is started on its course. *The combustion of the fuel heating the air tends to maintain the pressure which otherwise is constantly decreased by the expenditure in driving the engine.*" Thus, in 1872, Victor von Scheliha was enunciating the heater principle, an aspect of torpedo development which was not exploited by the major manufacturers until its rediscovery in 1901, when a Woolwich technician found, quite by chance, that the warming effect of sea water on the compressed air propellant produced an increase in speed of half a knot.[14] It was a seemingly insig-

nificant amount but it opened the door to a new way for improving performance, and with the Anglo-German naval race about to begin there was a mad rush to achieve higher speeds and longer ranges. Sir W. G. Armstrong Whitworth and Company demonstrated their first Elswick heater in 1905, having taken out patents the previous year. And in 1906, the Whitehead company produced its own system known as the "wet" heater. Finally, in 1908, the Royal Gun Factory, Woolwich, introduced a simplified heater based on the work of Lieutenant Hardcastle.[15]

In fact, it had taken the world's most experienced and knowledgeable torpedo designers more than thirty years to exploit von Scheliha's inspirational flash of genius. Yet so far as can be established, the rebirth of the heater principle came from men who had never seen the original 1872 patent and had most certainly never heard of Victor von Scheliha. It is understandable, therefore, why this obscure German military railway engineer should be regarded as an inventor almost equal in stature to Robert Whitehead, John Ericsson, Louis Brennan,[16] and John Howell.

Von Scheliha may have been the first to heat compressed air to improve performance, but in fact, a number of later inventors also devised their own methods of exploiting the heater principle before the major manufacturers followed suit at the beginning of the twentieth century, and details of their various systems will be examined at the relevant points of the chronology.

There was, however, one glaring defect in von Scheliha's torpedo. There was no mention of a warhead in the provisional patent specification, and a torpedo which could not detonate an explosive charge beneath the target's waterline served little purpose, no matter how technically brilliant its design. The subsequent full patent of 1873 explains this apparent omission, for, as will be seen later, it makes it clear that the earlier version was only the carrier vehicle, a type of unmanned remotely controlled submarine, which was to be used to take the explosive charges or "torpedoes" to the target before launching them against the enemy.[17] Unfortunately, the colonel's use of the word "torpedo" when referring to the delivery vehicle instead of applying it to the explosive charges carried by, and launched from, the former does not result in clarity.

But leaving this small confusion aside, it must be said that the weapon considered by the Admiralty Torpedo Committee was not identical to the device described above:

The St. Petersburg Connection · 33

Early in 1871 Colonel von Scheliha proposed a locomotive Automatic Torpedo, similarly controlled and of similar dimensions and method of propulsion to that just described [this was a reference to the Lay torpedo which will be examined in detail in chapter 3] the motive power being carbonic acid developed by the action of sulphuric acid upon bicarbonate of soda; this Torpedo was provided with air bags, by which means it was intended to be submerged at any desired moment before its explosion. This proposal was subsequently extended to the construction of an automatic locomotive plunging boat, containing a series of torpedoes arranged so as to be separately released for the destruction of a whole fleet; in this arrangement the turbine system of propulsion was substituted for that of the screw. The estimated cost of this Torpedo Boat [*sic*] was not stated.

The committee had been quick to realize that the "torpedo" was a plunging boat and the explosive charges, torpedoes. But in clarifying this aspect of the weapon, the committee had added two more puzzles. Reference was made to carbonic acid–gas propulsion, yet on line fifteen of the 1872 provisional patent, von Scheliha specifically states, "The compressed air works an engine." Neither does the specification refer to a "turbine system," as reported by the committee, for line 17 of the patent document states that the "engine acts directly upon the shaft of a screw." The 1873 patent does not resolve the problem either, for it says very simply and unambiguously that "the compressed air . . . works an engine." And two lines later, "The engine acts directly upon the shaft of a screw." From this only one conclusion can be drawn. The "torpedo" examined by the Admiralty committee was an improvement on the weapon described in the 1873 patent using carbonic acid–gas propulsion and some form of turbine in place of the screw.

Generally speaking, much of the specification included in the 1873 patent is identical in wording to the earlier 1872 provisional patent, although, on occasion, new details emerge. For example, "The motor is air-compressed to many (say 30) atmospheres in an air reservoir forming part of the torpedo." By the time of the 1870 Sheerness trials, Whitehead's torpedo was operating on an air pressure of 1,200 pounds per square inch. Von Scheliha's estimated 450 pounds per square inch was therefore relatively modest and certainly achievable.

The patent specification and its appended large-scale drawings also confirm the committee's description of what can only be called the mul-

tiple warhead scheme which von Scheliha had devised. Having explained how the operator could submerge the "torpedo" by means of an electric current together with a system of air-bags and "taps" (presumably valves), the specification continues: "The torpedo carries on its back a number of charges, each in a separate case, and these can be detached at the will of the operator. Each charge case or shell as soon as it is set free floats to the surface of the water but is still connected with the torpedo by an insulated wire, and an electric shock [pulse] can at the will of the operator be transmitted through this wire for the purpose of exploding the charge. The charge case or shell is also furnished with a percussion fuse [presumably some form of contact detonator] which is unlocked for action at the time the charge case or shell is detached from the torpedo." Once again the confusing terminology does not make for easy interpretation. The so-called torpedo appears to have been a small remotely controlled submersible which, having been steered to the target on the surface by the operator, is submerged in the final moments to take it under the enemy's keel. Released by the operator, the explosive charges, the *real* "torpedoes," then rise upward to detonate on striking the hull of the target or, alternatively, are detonated electrically.

In a later section of the specification, von Scheliha stated that the charges contained fifty pounds of dynamite while the drawings indicate that they were released by some form of spring action actuated by electricity. Unfortunately, there is no detailed drawing of the explosive charge itself and so there is no satisfactory indication of the air chamber's dimensions or capacity. Whether its volume would have been sufficient to overcome the weight of the warhead and the surrounding steel casing and achieve the positive buoyancy necessary for the device to float freely upward is problematical.

This new "torpedo" seems to have been thoroughly impractical and, to be brutally honest, a typical Victorian monstrosity which sadly seriously dents the inventor's credibility. Had he been content to remain within the limits set by the 1872 weapon, which Wellesley saw demonstrated on the Neva and confirmed "were most satisfactory," perhaps von Scheliha's story may have had a happier ending. But it seems that, by this stage, he had ventured into uncharted waters he was unable to navigate. Having examined every other aspect of his formidable torpedo in minute detail, he shrugs off the final crucial moments of the attack: "Each charge case

Col. Victor von Scheliha built and demonstrated the first practical dirigible weapon. This torpedo carried explosive charges on the upper part of its body with the optimistic aim of sinking an entire fleet.

. . . as soon as it is set free, floats to the surface," adding vaguely, "And being by an air chamber rendered lighter than the volume of the water it displaces," it will rise to the surface. Von Scheliha's Teutonic origins are apparent from his syntax. His cavalier optimism, however, suggests that he was living on another planet.

It is difficult to know why the colonel embraced the multiwarhead concept torpedo, although in itself the idea was a tribute to his farsightedness. And having designed a submersible that was both practical and workable, why was he not content, like Whitehead and Ericsson, to allow the device to strike the target and explode its charge by means of a simple impact trigger? A bird in the hand was, surely, better than two or more in the bush. But as Lord Fisher's aphorism puts it, "While striving for the gnat of perfection, we swallow the camel of unreadiness." Thus von Scheliha, by seeking to sink several ships at one and the same time, missed a golden opportunity to pose a genuine threat to Robert Whitehead's monopoly.

Although we do not know the ultimate end of the colonel's story, a few snippets of information remain: a provisional patent application for a system of electric railway signals submitted on 23 November 1875,[18] which gives his military rank as a lieutenant colonel and shows him as "temporarily residing" in St. Petersburg, and a letter dated 9 October 1874[19] in which he criticizes an article published in the *Times* on "Offensive Torpedoes in which my name is mentioned."[20]

This letter fortunately throws some further light on the weapon referred to in the Torpedo Committee report. Having first put the case for controlled torpedoes as opposed to the Whitehead free-running weapon, the latter being dismissed as "to be as futile as for a boy having set the

rudder of his toy ship to expect it to reach any given point on the banks of the Serpentine," the colonel went on to say that "the motive power by which my torpedo boat is propelled is gradually generated in the boat itself while running thereby obviating the necessity of a large reservoir subjected to high pressure." This is a reference to the carbonic acid–gas propellant examined by the Torpedo Committee and mentioned in its report. He then continued, "The screw, being apt to foul in nets or rope obstructions, is replaced by a more suitable propeller." This is, undoubtedly, a reference to the turbine which was also referred to in the report and, in context, it is clear that von Scheliha had only adopted another form of propeller and not a turbine propulsion scheme.

In a final flourish, he describes employing a battery of no fewer than *twelve* "heavily-charged torpedoes which . . . may be discharged one after the other at the will of the operator." The gargantuan sea monster was apparently growing with every passing year, for although the 1873 patent had referred to "a number of charges," the accompanying drawings showed a maximum of only eight in place. The patent specification, however, added, "The number of charges carried by any one torpedo may be considerably increased by attaching tubes to each side of the torpedo into which [further] charges are placed." When considering the weapon as a whole, it must be remembered that twelve charges, based on the total weight of the explosives alone and taking no account of the steel casing, would add a massive six hundred pounds to the weapon—a very different proposition from the unarmed torpedo that had been successfully tested on the Neva—and one which would have probably transformed that triumphant demonstration into a humiliating failure.

There is a curious coda concerning von Scheliha's fate in Wellesley's memoirs.[21] As explained earlier, the Russians had encouraged the colonel to move into an expensive hotel so that he could suitably entertain senior officials while the torpedo was being built. "The result was very different from what he had anticipated," Wellesley noted in his journal. "The [Russian] Admiralty, once in possession of his plans, and requiring nothing more of him, informed him that the experiments did not justify the purchase of his invention. At the same time the hotel keeper and other tradespeople began to press for payment of their accounts which the wretched German colonel had fondly hoped he would be able to settle when paid by the Russian government for his invention."

Wellesley said that the inventor's constant visits ceased abruptly and the last time he saw Victor von Scheliha, the latter told him he had been imprisoned for debt. This strange meeting occurred while von Scheliha was driving through the streets of St. Petersburg in a *droshky* with a companion. Having stopped to speak to the English diplomat, the colonel explained that the Russian penal system allowed prisoners a few hours of liberty on license providing they were accompanied by one of their creditors who were required to stand surety for their safe return to prison. Wellesley never saw him again.

It was a sad end for the man who at one time seemed set fair to challenge Whitehead for supremacy in the torpedo market and yet ultimately sank into such obscurity that his name and his weapon have never before been mentioned in any known book or article on the development of the torpedo. So far as history is concerned, Col. Victor von Scheliha and the St. Petersburg connection had turned out to be a blind alley.

Soviet propagandists used to assert that the Russian engineer Ivan F. Alexandrovsky–another resident of St. Petersburg and a contemporary of von Scheliha–invented the self-propelled torpedo in 1865, several years before Robert Whitehead. It was a difficult claim for Western historians to challenge without access to the relevant archives. And their task was made no easier by the complexities of the Slavonic language and the problems posed by the Cyrillic alphabet which makes it difficult to even scan a page rapidly in search of key words or even initials. But facts that are now coming to the surface in Russian technical literature scarcely support the former communist regime's attempt to usurp Robert Whitehead's contribution to the torpedo in its desire to rewrite history.

Ivan Alexandrovsky was born in 1817 and is alleged to have informed the Russian Admiralty in 1865 that he had devised a "self-propelled mine" capable of reaching ships and sinking them. Bearing in mind that mines were known as torpedoes at that time, there was obviously plenty of room for confusion, particularly by nonspecialists, although admittedly, the "self-propelled" prefix would suggest some form of fish torpedo. The facts become more clear in a document dated 20 November 1868 in which Alexandrovsky listed out several early projects he had previously submitted to the czarist authorities. Among these was a submersible armed with "floating mines."[22] The use of this latter term implied that his "torpedo" was no more than a "drifting mine," a weapon that merely

moved with the tide or current. It was to be delivered to the scene of operations by a submersible in much the same way that Fulton's mines were used at Boulogne in 1804, when they were towed into action by rowing boats before being cut loose and left to the vagaries of wind and tide.[23] Thus this early written evidence also fails to corroborate the claim that Alexandrovsky was the first in the field, for Whitehead had been working on the concept since 1864, when he had initially met de Luppis. And as his weapon was ready for testing by the end of 1866 it is reasonable to assume that it was well on the way to completion the previous year.

To quote Alexandrovsky's own words, "In 1869 I made a representation to Vice Admiral Popov for a project just invented by myself—a torpedo.[24] I made a request for funding so that I could explore its possibilities and to develop it." It is apparent from this that, even in 1869, his torpedo was little more than a paper project, whereas, by then, Whitehead's weapon had already been successfully tested and sold to the Austrian government in August 1868.[25] These two facts alone are sufficient to demolish the old Soviet claim that Alexandrovsky had invented the self-propelled torpedo before Robert Whitehead and this aspect need be pursued no further; however, as much of this early work in St. Petersburg is unknown in the West, the story of Alexandrovsky's weapon will be of undoubted interest to enthusiasts.

His involvement with the torpedo, despite the 1865 meeting with H. K. Krabb, a senior Russian Admiralty official, was in fact triggered by the naval ministry itself when details of the successful Austrian trials of the Whitehead were received in St. Petersburg in 1869 and he was invited to build a similar weapon. This had led to his approach to Vice Admiral Popov in search of funds. In the course of these negotiations, one of his preliminary drawings was presented to Grand Duke Constantine Nickolaevitch, whose intervention resulted in the Admiralty agreeing that, while Alexandrovsky would have to build the torpedoes at his own expense, he would be guaranteed governmental financial credits.

Working from private workshops in Kazan Street, St. Petersburg, it took the inventor five years to produce his first two prototypes. Both were cigar-shaped with a slightly blunted nose, unlike the needle bows of the contemporary Fiume torpedo, and were made from 3.2-mm steel sheeting. Their diameter was twenty-two inches with a length of twenty-four feet and a weight of 2,425 pounds. The compressed air propellant was

The St. Petersburg Connection · 39

stored at sixty atmospheres, but this pressure was reduced to five or ten atmospheres before being fed into the single-cylinder, double-acting piston engine which was directly coupled to the shaft of a single propeller at the stern.

Trials of the new weapons took place in the summer of 1874 near Kronstadt over a range of about two thousand yards. Depth-keeping proved to be satisfactory, but the speed was disappointing, averaging just over five knots. Indeed, the official report on the trials referred to the torpedo as "cumbersome" and the speed as "mediocre." The Russian Admiralty demanded a faster weapon and gave Alexandrovsky a deadline of 15 August 1875 to come up with a result, although this time the Navy agreed to build the improved model in its own workshops.

The inventor was doubtful of achieving success in such a limited amount of time, but he hopefully fitted a new two-cylinder engine and kept his fingers crossed. The additional tests in the spring of 1875 witnessed an increase in speed to 13.8 miles per hour. This improvement, however, was only achieved at the price of erratic depth-keeping, and Alexandrovsky was forced to try a new depth-control mechanism consisting of two horizontal wheels combined with hydrostatic valves. He claimed that this device was "nothing more than a miniature copy [of the apparatus used in] the submarine invented by [me]." This latter vessel, as already noted, was alleged to have been built before 1868.

Duly modified, the new weapon was completed in late 1875, but adverse weather delayed practical tests until 1876. In its improved form, the torpedo's diameter had been increased to exactly twenty-four inches and the body shortened to nineteen feet, although both the weight and shape remained unchanged. No details of the 1876 tests have been traced, but it was admitted by the Russian Admiralty that it was "disappointed in Alexandrovsky's submarine, its torpedo, and Alexandrovsky himself." The reason for this disappointment was not stated. It may, in fact, have been no more than a political smokescreen, for on 11 March 1876, the czar's government had signed a contract with Robert Whitehead's company for the supply of one hundred 15-inch torpedoes, delivery of which actually began on 29 May 1876, only a matter of weeks after the tests of the Alexandrovsky weapon, with the arrival in Russia of the first batch of twenty torpedoes with the serial numbers 171–190. Significantly, the Fiume-built torpedoes were not only better made, more reliable, and handier in

size and weight but also much cheaper. And cost savings loomed large in the Imperial Navy's budget.

In 1878, another version of the Alexandrovsky torpedo reached twenty and a half knots, a speed that was edged up to twenty-three knots two years later and was only slightly slower than the contemporary Whitehead 15-inch. But it was by then too late to save the day, and the inventor had to call a halt and accept defeat after his prolonged struggle with the all-conquering Whitehead weapon. Ivan Alexandrovsky died in 1894.

Another St. Petersburg inventor, Col. Ottomor B. Gern, can lay claim to have produced the largest and heaviest torpedo ever built. He also beat Alexandrovsky to the post by unveiling it to the public in 1872, two years before the Kronstadt trials of the latter's first practical prototype. Weighing six tons with a diameter of thirty-nine and a half inches, it was somewhat surprisingly only a modest twenty-three feet in length and was intended to be carried underneath the submarine, which the colonel had designed and built in 1867. It was fitted with a compressed-air engine and incorporated depth-control gear similar to that employed by Alexandrovsky, with two horizontal wheels operated by hydrostatic valves. Horizontal stabilizers in the form of external fins were provided to keep the weapon on a straight course. This monster was tested in 1872 but showed "no more than modest results" and, like so many others, was consigned to oblivion.

Colonel Shapovsky's underwater rocket, in which petroleum spirit was used as a propellant, is the final pioneer Russian torpedo for which information can be traced. It was notable for its gyroscopic stabilizing mechanism—the first example of this particular technology being applied to torpedoes—and its use by Shapovsky places him more than ten years ahead of Whitehead, whose first experiments with the Russian Petrovitch gyroscope were not carried out until the last decade of the century. Rated as "the best example of all models of self-propelled underwater mines [*sic*]" by the Russian Admiralty, it nevertheless failed to live up to this assessment when it was tested in 1879. The fuel burned evenly above the surface, but it proved to be a very different story when the weapon submerged, for it then promptly exploded and destroyed the rear end of the device.

Let us hope that one day, when archives become more freely available and computers have overcome the problems of translation, a treasure trove of other Russian weapons will be placed in the public domain.[26]

CHAPTER 3

Extreme Simplicity

A RESPECTED torpedo historian once said of John Louis Lay, "It is inventors like him that help to avoid dullness."[1] An examination of his
somewhat erratic career would seem to support that observation.

Born in Buffalo on 14 January 1832, Lay joined the United States Navy
in 1861 as a second assistant engineer and two years later was promoted
to the rank of first assistant engineer. His flair for engineering and mechanics had been noted while he was still at school, and his aptitude for
converting discarded boilers into torpedoes—at that period the term also
applied to static underwater explosive devices which would, today, be
classified as sea mines—soon attracted the attention of his senior officers.

In September 1863, when the Union government decided to play the
Confederates at their own game and deploy underwater weapons against
their enemy, the device they chose to send into battle was a spar torpedo
invented by the young John Lay and Chief Engineer William W. W. Wood.
Author Peter Bethell was of the opinion that Wood made little contribution to the weapon and was probably only a party to the invention by reason of his senior rank, which, he implied, could be used to open influential doors. In his patent, however, Lay specifically credits the design of the
shells employed with the spar to "Chief Engineer W. W. W. Wood and myself," so it is possible that Wood was the ordnance expert while Lay was
concerned more with the mechanics of the spar torpedo's operation.[2]

Just as there was powerful opposition to the Whitehead torpedo in
1870, numerous voices at the highest levels of command in the Union
navy condemned the whole concept of underwater weapons. And even
Rear Adm. David Glasgow Farragut, famous for his defiant "damn the
torpedoes," proclaimed his opinion that the use of such devices was "un-

[41]

worthy of a chivalrous nation," although he accepted that they would prevent giving "your enemy such a decided superiority over you."[5] Or, to put it another way: if you can't beat 'em, join 'em.

Lt. William Barker Cushing's attack on the Confederate ram *Albermarle* using Lay's weapon earned the young officer a hero's place in American history, and full details of his exploits will be found in the chapter on the spar torpedo. Not surprisingly, the success of his new weapon on its first outing under combat conditions brought John Lay a sense of pride in his achievement which led him to spending the rest of his life in search of bigger and better torpedoes. Bethell, perhaps a little unkindly, criticized Lay's weapon on several grounds and dismissed it as "simply a spar torpedo fitted to a steam launch; although the use of any word connected with 'simple' is hardly justified [when referring to] a Lay torpedo."[4] Having examined all four patents taken out by Lay and Wood for this particular weapon, it is difficult to disagree with his assessment.[5] For in trying to achieve perfection, John Lay, like Colonel von Scheliha, pursued his objective with mind-numbing detail. This brief extract, concerned with determining the angle at which the spar should enter the water, is, of course, related to the detailed drawings included with the patent application. But its complexity is self-evident bearing in mind the basically simple nature of the operation involved:

> If it be necessary to move the operating bar H so as to be at a greater incli-
> nation than that shown in *Fig. 1*, the attendant so manipulates the valve-
> rod y of the steam cylinder x' that the steam will act on the face of the pis-
> ton x''' and cause the piston-rod to pull the chain u'' which, through the
> arm X''', shaft X', arms X X, and chains W''', elevates the cross-head V,
> and with it the plates U'', U', the sleeve T, and operating bar H, and other
> parts connected therewith, the bar turning on a center coinciding with the
> center of the sphere S, which, together with the plates w and w' and stuff-
> ing-box N, form a perfectly watertight ball-and-socket joint.[6]

And all of this in just one sentence. Lay clearly did not believing in stopping to draw breath. However, so far as the attack on the *Albermarle* is concerned, Commander Bethell deserves the last word: "Nevertheless [the device] worked. Which is more than can be said of most of Lay's torpedoes."[7]

Extreme Simplicity · 43

Before returning to the strict chronology of the automobile torpedo's development, this would seem to be an appropriate point to follow the peaks and troughs of John Lay's career until his death in 1899.

Having resigned from the U.S. Navy at the end of the Civil War, he was approached by the Peruvian government to organize the underwater defenses of Callao Harbor, which was facing the threat of an attack by the Spanish fleet. The work was accomplished by 1867, and on completion of his contract, Lay returned to Buffalo, where he began the task of building his new "movable torpedo submarine." The confusion in terms was apt, for, like von Scheliha's weapon, the main part of the device was no more than an awash-running delivery vehicle.

According to the Admiralty Torpedo Committee–which considered the weapon along with those of Ericsson, Whitehead, Harvey, and von Scheliha–Lay claimed that his torpedo was superior to the Whitehead because its movements remained under control after it had been launched and "the errors in its course caused by tide or other unavoidable circumstances, could be rectified at will by electric agency through a cable connecting the torpedo with the post of observation."[8]

The committee noted that the Lay torpedo had first been tested at New York in November 1872, though the results of these American trials were not disclosed. It was described as an iron boat, twenty-three feet in length and three feet three inches in breadth and similar in shape to the Whitehead, though the latter statement is not supported either by the drawings in Ericsson's article on the weapon[9] or, indeed, by the term "breadth" in the committee's reference to dimensions, which suggest it was boat-rather than cigar-shaped. Divided into five compartments, it was driven by a four-bladed propeller powered by a carbonic acid gas–fueled engine. The bow compartment held a 500-pound nitroglycerine explosive charge, considerably bigger than the warhead of the contemporary Fiume 16-inch standard model, which weighed a maximum of 117 pounds.

As Ericsson was quick to point out, the additional weight of explosives was necessary simply because the Lay weapon struck its target at the waterline where armor protection was often at its thickest. The Whitehead, of course, exploded against the unprotected hull below the waterline where even a small degree of damage could be fatal. Immediately behind the warhead came the compartment used to store the carbonic acid pro-

pellant, while the third space contained two and a half miles of electrical cable coiled around a wheel through which signals were transmitted for the purposes of steering corrections. The two remaining compartments in the stern section provided space for the batteries and motors that "controlled the screw and tiller." Its speed was claimed as eight miles per hour with a range of two miles.

The Lay torpedo's major disadvantage, however, rested on the fact that it was really a surface weapon which ran awash. Indeed, apart from the use of electricity instead of tiller lines, it showed little advance on the original "coastal fireship," which had been built nearly ten years earlier. As a result, and as already noted, the explosive charge detonated too far above the waterline to compromise the buoyancy of the target vessel. Appreciating the problem Lay subsequently modified his initial design before the weapon was submitted to the Torpedo Committee by S. D. Caldwell, who, by that time, had apparently acquired the rights from the inventor. According to the Committee's report, "The head containing the explosive charge became detached upon its coming into contact with the hull of the [target] ship . . . and dropped down to a depth of seven feet before exploding."[10] It was, to say the least, a somewhat strange solution.

But Ericsson had already evolved a scheme which produced a similar result, lowering the explosive charge from an awash boat by means of pivoting spars as it closed with the target.[11] It does not seem to have been developed, so it presumably failed to live up to its inventor's expectations. Without illustrations or plans of the system adopted by John Lay, it is impossible to compare it with that of his rival. But a review of early torpedo patents produced the Mills Callender weapon of 1863, already described in chapter 1, in which the charge swung clear of the "torpedo" on impact and, attached to a chain, swung in an arc beneath the keel of the target vessel. This action would seem to match the committee's observations and one wonders whether Lay "borrowed" the idea from Callender.

Significantly, the Royal Navy's experts were only shown the plans of the modified weapon rather than the torpedo itself, although the committee noted that "it was after[ward] experimented with on several occasions by the officers of the US Navy, as detailed in the reports of the British naval attache in Washington."[12]

One of the American officers concerned, Cdr. William A. Kirkland, earned Ericsson's wrath by making a favorable comparison of Lay's

Extreme Simplicity · 45

weapon with the former's pneumatic torpedo. And in a two-part series of articles in *Engineering*, he set out "to examine the leading features of the [Lay torpedo boat] . . . and at the same time correct the erroneous calculation and serious mistakes of Commander Kirkland,[13] regarding my movable submarine torpedo."[14] The public squabble between Ericsson and Kirkland are of little interest to the torpedo story but, in the course of the article, hitherto unknown details of the Lay weapon emerged which demonstrate that his technical thinking was far more advanced than that of his Swedish-born rival.

According to Kirkland, "The [propellant] gas is carried in liquid form. There is enough of it to drive the boat two miles. When it expands as vapour, a great loss of temperature is the result, and this might diminish the pressure greatly. This evil has been experienced in other carbonic acid motors, in some of which the volatisation of a part of the liquid froze the remainder, causing the pressure to cease altogether. It is counteracted in [the Lay torpedo] partly by the larger size of the wrought-iron reservoir or flask, [and] partly by . . . conducting the gas past the reel compartment to the reducers through small pipes running along the outside of the [body] shell, and thus exposing a large surface to the water, which imparts some heat to the gas within."[15] Thus Lay, like von Scheliha, had appreciated the effect of warming gas (or compressed air) by bringing it into contact with sea water, a concept not adopted by the major manufacturers until the effect was observed by a Woolwich technician in 1901.[16]

Ericsson, however, chose to ignore the significance of Lay's discovery–or, less likely, failed to appreciate the phenomenon–and railed against the complications of the rival weapon. "Two distinct motors are necessary to start the third motor which actuates the propeller; that other motors are required to operator the rudder, and that in order to make good the weight lost by paying out the [electrical] cable, an 'adjusted cock' has been devised for admitting water into the boat, opening while the boat is moving and closing when [it] stands still (a great mechanical achievement, experts will admit, provided this cock infallibly turns the right way at the right time)."[17] He also criticized the time-consuming and minutely detailed procedures necessary to refuel the torpedo following a test run, although he seemed to have overlooked the fact that, in operational use, the weapon would not be returning for a second attack.

46 · *Nineteenth-Century Torpedoes and Their Inventors*

"Commander Kirkland cannot perceive that the mechanism thus described is at all complicated. In fact he says 'its great value is in its extreme simplicity,'" Ericsson thundered.[18] Apparently lost for words, he threw down his pen and said no more. In the circumstances, perhaps the last comment should rest with the Admiralty Torpedo Committee, which noted dismissively, "The cost of [a Lay] torpedo was estimated at £2,200."[19]

With the contemporary Fiume-built Whitehead Standard 16-inch torpedoes being offered for sale to customers at three hundred pounds apiece, John Lay's unproved weapon was clearly not much of a bargain.[20]

Although Lay took out six separate patents for steam engines and another for a railway locomotive following his return to the United States in 1867, there is little doubt perfecting his torpedo became the dominant driving force of his life thereafter. But his continual tinkering with the original basic design, usually by making its internal workings progressively more complex, has led to a great deal of confusion. And many authorities provide identical performance and specification details to all of the various models he built—from the 1872 version already examined on previous pages to his dirigible weapon of 1887.

Before describing the Lay torpedo's first unhappy experience of combat during the Chile-Peru war of 1879–82, it is fortunately possible to provide precise details of the weapon involved which have been taken from a report in the *Engineer* in a report dated only seven months after the errant torpedo had caused so much alarm among its Peruvian friends.[21] The report in question also included detailed scale plans, and the proximity of its publication with the events in South America suggests that the text and drawings almost certainly relate to the torpedo fired from the Peruvian ironclad *Huascar* on the night of 27 August 1879.

The length of the weapon was still given as twenty-three feet, identical to the original 1872 model, and it is this similarity that has led to the many misleading descriptions of Lay's various weapons. It should be noted that the details given in this chapter have been taken either from U.S. patent specifications or from reports in responsible professional journals of the relevant period. If there is any doubt as to the correctness of the data it is indicated in the text. Thus although several writers credit the length of the Lay-Haight 1894 dirigible torpedo as also being twenty-three feet, contemporary photographs suggest that it was considerably

Extreme Simplicity · 47

larger. What remains in dispute, however, is the date of this gargantuan weapon which has been shown variously as 1880, 1883, and 1894.

As already noted, the Torpedo Committee report described the 1872 model as "similar in shape to the Whitehead" but then promptly contradicted itself by referring to its "breadth" of three feet three inches.

Ericsson's drawing in *Engineering,* however, depicts it as cylindrical and pointed at each end. He gives its length as twenty-five feet with a diameter of three feet.[22] One could be forgiven for thinking that the Torpedo Committee's members and John Ericsson had been looking at two totally different weapons. But, unfortunately, this is one of the recurring problems with the Lay torpedo. From all this confusion, however, it seems reasonable to deduce that the 1872 model was a gargantuan cigar-shaped weapon with a diameter of between three feet and three feet three inches, with an overall length of twenty-three to twenty-five feet.

But to return to the report in the *Engineer* relating to the Peruvian torpedo, the diameter of this particular weapon is not stated in the text, but the detailed scale drawings show it to have been approximately twenty-four inches, based on an assumed overall length of twenty-three feet. It was built from "thin plate iron or steel" and its 9-IHP (indicated horsepower) engine was powered by "the expansive force of carbonic acid gas." An alternative engine running on ammoniacal gas was available, however, and this is, perhaps, why there are often references to two versions of this torpedo. The propeller system, too, was offered in three variations: single screw, double screw, or two screws. In the two-screw model the propellers were of contra-rotating type working on a three-wheel bevel system similar to that originally adopted by the Royal Laboratory, Woolwich, the fascinating story of which will be examined in the next chapter.

It is clear from the descriptions that the torpedo was intended to operate submerged or partially submerged, an obvious improvement on the 1872 version. In view of what happened after the weapon was fired the Peruvian torpedo was almost certainly semisubmersible in operation. The "boat" was provided with "a double set of side wings or horizontal rudders."[23] However, it is noted that the angle of these had to be adjusted externally before the torpedo was launched. Two guide rods which could be raised or lowered by electrical remote control so that the operator could determine the position of the weapon while running were also

fitted. For, as with the 1872 version, the most important characteristic of the torpedo was its ability to be controlled remotely via an electrical cable. But like all of Lay's inventions, the details of how this system worked are too complicated for a nontechnical textual description. And matters are made no clearer in the drawings as the crucial control keyboard is not illustrated. For those who enjoy unraveling Chinese puzzles, the relevant part of the text relating to the control system reads:

One end of the cable is connected to a keyboard at the station on shore or on board the ship. . . . This keyboard is provided with a suitable battery or other means of generating the electric current. The cable is composed of several wires, each of which is insulated from the others. One of these wires is connected with the mechanism for starting and stopping the boat, one is connected to the steering apparatus, one serves for indicating to the operator at all times the exact position of the rudder, one is connected with mechanism for elevating and depressing the guide rods, and serves for firing the charge in the magazine. The motive power for effecting the necessary movements of the mechanism or apparatus in performing the above operations is obtained from the engines, which are provided with suitable valves arranged in combination with electromagnets, shunts, and devices connected with the wires of the cable.[24]

One suspects that the journalist writing this piece was as bemused as he sounds. And Kirkland's phrase "extreme simplicity" is not the one that immediately springs to mind. However, the *Engineer* report stated that "experiments made by the Russian government with the Lay have been productive of such satisfactory results as to enable the inventor to dispose of his Russian patent."[25] It is to be hoped that John Lay was more successful than Colonel von Scheliha in obtaining the purchase money from his St. Petersburg customers. According to a subsequent section of the report, "The Russian government, in addition to the purchase money, has subsidized Colonel [sic] Lay to erect works in Russia for the manufacture of his weapon; the necessary plant, machinery, and skilled labour having been imported from the United States."[26] Ten of the torpedoes were said to be already in the course of construction.

Many of the features examined earlier in this chapter were incorporated into the patents which Lay obtained on 14 January 1879, and these appear to have been taken out with a view to protecting various devices

Extreme Simplicity · 49

and arrangements which he had been using since at least 1872.[27] As the patent application was lodged on 18 December 1878, this sudden rush for legal protection of his ideas was probably connected with his promotional sales visit to Europe, which had begun a short while earlier. Unfortunately, the plans attached to those patents are misleading when it comes to determining the weapon's outward appearance and form. For Lay notes that the drawings of the torpedo (or torpedo boat as he describes it) are "relatively much shorter and of greater diameter than the boats usually made by me. This is for the convenience of illustration."[28] One is tempted to add "and the inconvenience of subsequent researchers." But on the basis of his remark, it must be concluded that the actual dirigible torpedo must have been more cigar-shaped than the patents imply. And, certainly, if the proportions and shape of the weapon had reflected those depicted in the drawings it would have been the object of considerable hilarity on the part of both the experts and the technical press.

Lay describes himself as "temporarily residing in St. Petersburg, Russia" at the time of the patent application on 18 December 1878, and this serves to confirm the report of his visit to Russia and his negotiations with the czar's government in the *Engineer*.[29] It is also apparent that the patented design formed the basis of Lay's later dirigible torpedo, the subject of the British and foreign trials to be examined later. St. Petersburg was, of course, Russia's seat of government, which explains Lay's presence in the city. But it was also the last known place of residence of Victor von Scheliha, and it is fascinating to speculate whether the two men met to discuss their respective weapons. There is, in fact, a significant clue to suggest that they could have been acquainted, for Lay's 1872 model employed an elementary "heater" system which passed the propellant gas through small tubes along the outside of the torpedo's body shell so that the warmth of the sea would heat it.[30]

In his 1872 patent, von Scheliha described his heater in unambiguous terms: "In some cases I arrange a small furnace within the compressed air reservoir and ignite the fuel at the time the torpedo is started on its course."[31] Lay, in his 1879 patent, adopted a similar concept of direct heat by passing the propellant gas through pipes which lead through a chamber containing alcohol. The latter was then ignited by an electrical spark generated by the current passing down the external control cable,

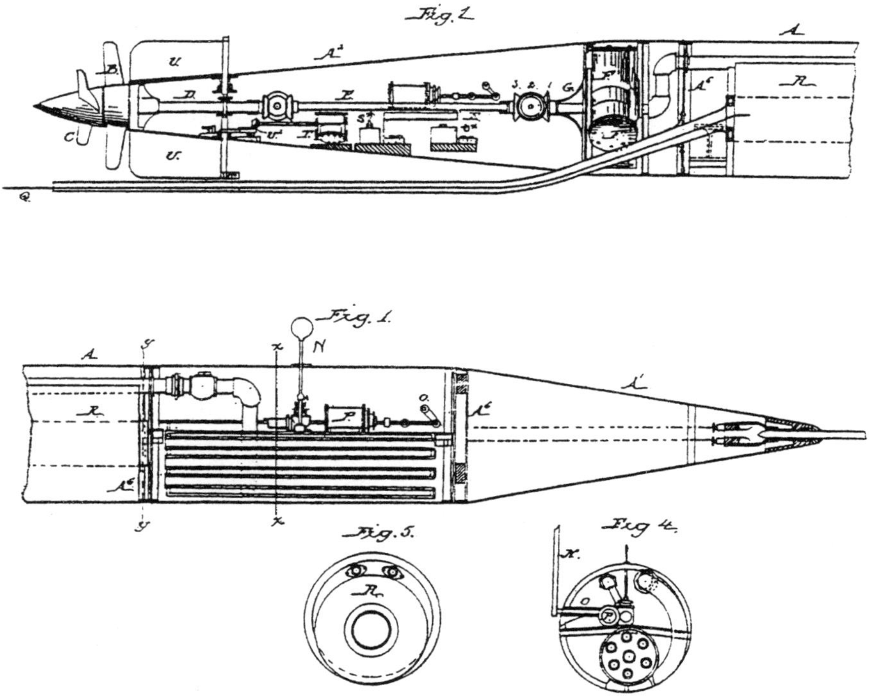

John Lay's 1879 torpedo, showing its correct proportions.

and the resultant naked flame heated the conductive pipes. The arrangement was virtually identical to the "small furnace" envisaged by von Scheliha in his specification. The abrupt change from relying on the warmth of sea water to heat the gas to employing the brute force of a naked flame to achieve the same object is so great that one is led to conclude that Lay may have been influenced by the German colonel's ideas, either by direct discussion with von Scheliha or by "borrowing" the concept after, perhaps, seeing the plans of the latter's weapon in the Russian naval arsenal. But whatever the truth may be, it is clear that both men were aware of the heater principle twenty to thirty years before it was rediscovered and applied to post-1901 torpedoes.

Oddly enough, exactly the same path of development was followed in the twentieth century. Having recognized the warming effect of seawater on compressed air and noted the subsequent improvement in performance, torpedo engineers had within a very few years moved on to the

Extreme Simplicity · 51

combustion chamber system to heat the weapon's air propellant.

Except for matters of detail, there was little further of interest in the 1879 patent. Among the refinements which Lay introduced was a braking system for the electrical cable together with a means of employing the propeller shaft to pay out the cable once the torpedo was up and running. In addition, and as yet one more complication, he provided two adjustable guide rods for the operator to steer by and, to add icing to the cake, incorporated into them rotatable discs together with lamps to make them visible during night operations.

The Russian venture, however, was not the limit of Lay's European sales drive. For on the night of 21 October 1879, a Lay torpedo was demonstrated on the River Scheldt at Antwerp in front of delegations from England, Belgium, France, Russia, Holland, Italy, Austria, Prussia, Denmark, Sweden, and the United States. The torpedo's speed was reported to have been nine knots, although it was pointed out that the new Russian models could make fourteen knots. Nevertheless, the demonstration was impressively successful when the torpedo passed safely through the twenty-foot gap separating the two moored barges on which the delegations were standing, after being steered across the Scheldt's powerful current for three thousand yards.[32] The plaudits at Antwerp, and the obvious satisfaction of the Russians in St. Petersburg, were in sharp contrast to the consternation that had greeted the Lay torpedo's first experience of combat conditions almost a year earlier off the Pacific coast of South America.

Hostilities had begun when Chilean troops occupied Antofagasta, an act of war which led the Bolivians to invoke their 1873 treaty of alliance and mutual support with Peru to bring the latter country into the conflict. The Peruvian navy, probably as a result of the contacts Lay had fostered while working on their harbor defenses from 1865 to 1867, had already purchased several examples of his dirigible torpedo, although some authorities have put the actual numbers as low as two. It has also been alleged that the threat of these weapons persuaded the Chileans to lift their close blockade of Iquique at dusk every evening for fear of being attacked by the Peruvians under cover of darkness. However, there seems to be little reliable evidence to support the claim, which might have been initiated by Lay himself.

No torpedoes were deployed during the skirmish between Peruvian

and Chilean naval forces which took place on 21 May 1879, when the ironclads *Huascar* and *Independencia* were challenged by Chile's *Esmeralda* and the gunboat *Covadonga*. A fierce battle ensued in the course of which *Huascar*'s guns sank *Esmeralda*, but Almirante Carlos Condell saved the day by using his seamanship to outmaneuver *Independencia* with his sole surviving gunboat, forcing the ironclad to run aground so that he could destroy the enemy at leisure with some well-directed shells. In that single action, the unfortunate Peruvians had lost half of their ironclad strength and *Huascar* found herself a lonely fugitive as Chile's warships tried to hunt her down.[33]

The next few months saw little by way of major actions, although *Huascar* made several brave but unsuccessful attempts to ram the enemy ships she encountered. Finally, toward the end of August, the ironclad's commander, Contra Almirante Don Miguel Grau, returned to Iquique and embarked two Lay torpedoes together with a trained operator. A few hours later he set out to find the enemy.

Grau's luck held, and on the night of 27 August, he located the Chilean corvette *Abtao* off Antofagasta and succeeded in closing to within two hundred yards of his quarry before being spotted. But instead of shaping course to ram his prey as the Chileans anticipated, Grau unexpectedly altered course so that the ironclad was broadside-on to the enemy and gave the order to launch a torpedo. The ungainly weapon splashed into the water like a novice diver and the electric cable paid out rapidly as the twenty-three-foot monster hurled itself at the enemy corvette. Excited by their success, *Huascar*'s crew lined the rails to cheer on the torpedo and watch it explode against the side of its hapless victim. But their excitement quickly turned to consternation and then fear as they saw the Lay suddenly reverse course and head back toward the ironclad like an avenging arrow. The operator tried to avert disaster by frantically pulling every control switch in sight and the crew's fear descended into blind panic when they realized that nothing could stop the runaway weapon.

Fortunately, one man, Teniente Diez Causeco, remained calm and, having carefully calculated the right moment, dived over the side and bodily pushed the torpedo off course so that it ran clear of the ironclad. Grau was not amused and, according to legend, took the two torpedoes ashore a few hours later and buried them in the cemetery at Iquique. To conclude the saga of *Huascar*, she met up with the Chilean battleship *Al-*

Extreme Simplicity · 53

mirante Cochrane, sailing in company with *Blanco Encalada,* off Angamos some six weeks later. Outnumbered and outgunned, the Peruvian admiral fought bravely and along with many of his men died in action. *Huascar* fought on until finally, reduced to just one single operable gun, she was forced to surrender.[34]

But the sorry saga of Peru's Lay torpedoes was not yet over, for on 3 January 1881, after the captured *Huascar* had been repaired and refitted at Valparaiso by the Chilean navy, she was attacked by a small Peruvian tug that had been equipped to operate a Lay weapon a few days earlier, a fact that suggests the Peruvians had more than two of the devices. But as on previous occasions, the torpedo failed to respond to the electrical impulses being transmitted through the cable by the operator and, in despair, the tug's skipper beached his vessel and destroyed the weapon in disgust.

Although no longer part of the John Lay story, *Huascar's* unique place in torpedo history needs to be placed on record. The armored turret ship, displacing 1,130 tons and armed with two 10-inch Armstrong muzzle-loading guns, had fallen into the hands of the rebel leader Nicholas de Pierola during the Peruvian civil war of 1877 and was branded as a pirate vessel.

The rebels were discovered by a British squadron under the command of Rear Adm. Sir Algernon de Horsey on the morning of 29 May while lying at anchor at the port of Ilo. Rejecting a call for his surrender for "piracy on the high seas," Pierola attempted to escape the British warships by exploiting the ironclad's shallow draft and steering a course through the shoaling waters of the Ilo estuary. But the corvette *Amethyst* cut off their line of retreat and, with the support of the frigate *Shah,* began to pound the rebel ship unmercifully with their guns. Still, by late afternoon, little serious damage had been inflicted–the Royal Navy's gunnery was never quite as good as its admirals thought it was, and in defiant desperation, de Pierola turned to ram the British flagship.

Captain Bedford, the *Shah's* commanding officer, decided to play his trump card and ordered one of the frigate's Whitehead torpedoes to be launched. It missed, and *Huascar* survived to fight another day. But the little ironclad had earned the distinction of being the first warship to be attacked by a locomotive torpedo. And, of course, two years later *Huascar* became the first vessel to launch a Lay torpedo in anger–and perversely,

54 · *Nineteenth-Century Torpedoes and Their Inventors*

the second to become the target of one. It is satisfying to record that the gallant *Huascar* was still in service with the Chilean navy as a guardship at the outbreak of World War II.[35]

Lay was apparently undeterred by the failure of his torpedo in South America, and it seems that most European navies were unaware of its unpredictable antics in the hands of the Peruvians, for, as already noted, although the tests in Russia and on the Scheldt were successful, the observers may not have been so enthusiastic had they known of the weapon's disastrous performance under combat conditions.

But Turkey, too, was now taking an interest in Colonel Lay's device, although during this period of European history anything the Russians favored was immediately taken up by the Ottoman government, and a series of tests held on the Bosphorus in December 1882 proved to be overwhelmingly satisfactory, an opinion shared by the noted torpedo skeptic, Hobart Pasha.[36] His assessment, published in an open letter to the *Times,* is of interest not least for his reservations about Lay's frequently wild claims.[37] And having voiced the criticisms of the naval and military observers at the Bosphorus trials that, at nine knots, the torpedo was too slow and that the proximity of the propellers to the surface rendered them vulnerable to defensive gunfire, he continued:

> Mr Lay undertakes under a heavy penalty to give 12 knots speed, and to immerse the screw sufficiently to prevent its being open to attack while in motion, in which case, I think Mr Lay will merit to be warmly congratulated on the great success which will inevitably result from his invention. Without any way interfering with the rapid advance of the Whitehead torpedo as a weapon of naval warfare, Mr Lay has now–supposing always that he is able to maintain his guarantees of speed and immersion–shown that in large rivers and estuaries, where currents and tides are strong and changeable as to speed, the Lay torpedo would be most efficacious, inasmuch as it can be guided unseen [through the water] and independent of any current.

In other words, it was inadequate for use at sea–a conclusion with which the Peruvians would have no doubt agreed–but was ideally suited as a defensive weapon where currents and tides rendered the Whitehead unsuitable. Hobart Pasha's appraisal was a far cry from Lay's boastful claims.

Extreme Simplicity · 55

The letter to the *Times,* however, inadvertently produces another puzzle, for Pasha refers to "Colonel Lay and his associate Mr Nordenfelt," while another report on the trials,[38] pointing out that Lay was demonstrating an older version of his weapon with a maximum speed of nine knots, added, "And [it] is not the torpedo now supplied by Messrs Nordenfelt and Lay." The only conclusion to be drawn from these two comments is that the manufacture of the Lay torpedo was now in the hands of the Swedish armament's tycoon, Thorsten Nordenfelt, who, as will be seen later, went on to produce his own torpedo in 1888. Unfortunately, no further details of this business association have so far come to light. It was, incidentally, the "improved" version of the torpedo that Nordenfelt and Lay supplied to Russia. This weapon had a claimed speed of twelve knots, and this increased velocity silenced one section of its critics by forcing the tail of the torpedo deeper into the water so that the screws were better protected from gunfire.[39] According to the report, this additional speed was achieved by "substituting two small screws [probably contra-rotating to judge by the plans] for the long-bladed [propeller] used in the [Bosphorus] experiments."

Bethell points out that Lay only succeeded in selling two of his torpedoes to the U.S. Navy, although he provides no dates for reference.[40] One of these was lost early in 1880 during tests at the Newport Torpedo Station. The reason given for the accident suggests that Lay was not only an inventor and a salesman but also something of a spin-doctor. "It appears . . . that this torpedo was under repairs at the time of its being launched by the officers passing out of their torpedo course at the American Naval Academy," he explained. "Several rivets were wanting in some of the bottom plates, and the water entering through the holes thus left open, naturally sank the torpedo."[41]

Lay subsequently achieved, and even exceeded, his claimed maximum speed of twelve and a half knots. According to a report in the *Times,* "On Saturday [29 September 1883] an interesting trial of the Lay torpedo took place at the works of the Patent Cotton Powder Company near Faversham. . . . [It was, in fact, staged in the estuary of the river Swale.] The torpedo was lowered into the water and at the word 'go' it went off with a great rush. It accomplished the first half mile in two minutes eighteen seconds . . . and after completing a mile it was brought back in good condition."[42] The time quoted for the half mile equates to a speed of thirteen knots.

The American inventor was persistent if nothing else, and in 1887, he persuaded Royal Navy and British Army observers to attend a demonstration to be held off Brightlingsea on Saturday, 4 March. The torpedo, basically similar to that purchased by Russia and a faster version of the model used for the Scheldt and Bosphorus trials, was described as being impelled by carbonic acid gas and steered by electricity. Its weight was given as one and a quarter tons (twenty-eight hundred pounds) and its length twenty-three feet. Somewhat mysteriously, it was said to have "the propeller in the head [and] it had a rudder astern."[43]

This curious feature creates a number of problems regarding the correct identity of the Brightlingsea test model. Bethell mentioned in a footnote, "One pattern of J. L. Lay's dirigible torpedo had its propeller mounted in a cavity on the underside [of the body], three feet from the nose, but this is the only such example known to the writer. Almost anything could happen to a Lay torpedo, and often did."[44] He carefully avoided dating this particular pattern but otherwise confirmed the feature as belonging to the dirigible torpedo.

Kirby accepted Bethell's statement but again gave no date. He, however, went further by suggesting that the bow propeller related to a different torpedo: "Still later we find the Lay-Haight weapon . . . these . . . had their propeller near the forward end of the hull, partially recessed to avoid damage."[45]

The Lay-Haight torpedo, March 1894.

Extreme Simplicity · 57

George E. Haight, a former employee of Lay whose name often appears as a witness on the latter's patent applications, was involved with at least two other inventors, William H. Wood and William E. Winsor. In his joint patent with Wood and Winsor, filed on 31 August 1883, he included a drawing of their torpedo which clearly shows the propeller at the stern in the conventional position.[46] Thus the Wood-Haight torpedo, and this appears to be the only version, did not have a recessed propeller in the bows. There is also little doubt that the weapon in the patent drawing is the same as that depicted in Kirby's article which he describes as an "early form of the Lay torpedo as built in the 1870s."[47] To confuse matters further, a U.S. Navy photograph of the Wood-Haight, reproduced in this book, appears to be identical to the weapon in Kirby's illustration. Yet it is captioned *Wood-Haight 1894*. This is not intended as a criticism of Kirby, whose work I hold in the highest regard, but it serves to emphasize the problems encountered by historians working in this hitherto underresearched period of torpedo development.

The definitive answer to the mystery of the Brightlingsea torpedo's identity was provided by C. W. Sleeman, a former lieutenant in the Royal Navy who was, by that time, serving with the Ottoman navy as a captain. He was also the author of *Torpedoes and Torpedo Warfare*, originally published in 1880, the only history of the torpedo to appear in book form until *The Devil's Device* in 1975.[48] As an expert of the period, he was privy to the correct facts, and in the torpedo section of the 1887 *Naval Annual*, he gave an accurate account of the Brightlingsea test model, which, in several respects, contradicted the report in the *Engineer*.[49]

Having given the diameter of the torpedo as eighteen inches, a detail that virtually everyone had neglected to mention, he made it clear that the demonstration featured one of Lay's weapons and not a Lay-Haight: "Mr J. L. Lay has recently devised a new pattern torpedo, in which the principle novelty consists in the placing of the propellers in a recess in the bow. This torpedo was officially tested at Brightlingsea in March 1887, but the performance was not successful."[50] Confirming that the power unit was a spherical engine Sleeman went on to reveal that the torpedo had, not one, but two propellers set into its bows. "It is provided with two propellers on a common shaft, one loose, and the other fixed thereto; the former is made to revolve by an interlocking boss, when the latter has made half a revolution. In each propeller the two blades,

58 · Nineteenth-Century Torpedoes and Their Inventors

instead of being on opposite sides of the shaft, are close together, the leading edge of one coming just behind the following edge of the other. Normally the two sets of blades are folded together, covering a sector of 70°, and stand upright within the recess in the bow where they work."[51] One cannot help asking why the *Engineer*'s reporter, who was present at the trials, having drawn attention to the "bow-mounted propeller" (and note the singularity), made absolutely no comment on the unusual blade configuration in addition to overlooking that the arrangement comprised two propellers.

Sleeman also cleared up the chronology of the Haight torpedoes: "The Lay-Haight was designed by Mr Haight . . . in 1881 and differs from the original Lay in drawing off the carbonic acid [propellant] in a liquid instead of [a] gaseous state, and in using a super-heater of sulphuric acid and lime. A speed of 15 knots was obtained with this weapon in 1883 . . . [and] it ran on the surface."[52] Sleeman's observations are supported by U.S. Patent No. 292,428, which Haight, together with Wood and Winsor, took out to secure protection for their heater system, which used the chemical reaction of sulfuric acid on lime to heat the carbonic acid propellant. The specification similarly refers to carbonic acid being held "in a compressed or liquefied state . . . before it enters the engine."[53]

Thus, from a chronological viewpoint it seems safe to say that the Lay-Haight dated from 1881, the Wood-Haight's patent was applied for on 31 August 1883, and that the only weapon with bow-mounted propellers was John Lay's modified torpedo, which had been developed from his earlier weapons.

Before finally moving on, a search of the U.S. Patent Office records revealed that Lay was not the first inventor to make use of a bow propeller on a torpedo. Henry F. Knapp of New York City applied for letters patent in respect of such an innovation on 23 September 1873.[54] According to the specification Knapp considered that a propeller placed "forward of the boat . . . [would] insure a straight and direct course," although he does not explain why this should be the case. The egg-shaped body was suspended from a surface float, and the engine was fueled by carbonic acid or steam. Steering was controlled by a strange spring-loaded helm which responded to compressed air supplied to the mechanism via a hose from an external pump. The oddest feature of the weapon was its impact-detonated explosive charge, which was attached to the forward end of the

Extreme Simplicity · 59

propeller shaft, the specification cautioning that it should be "set loosely so as not to revolve with the shaft." Henry Knapp, however, seemed to lack confidence in his bow propeller, for he observed, toward the end of the patent's text, that "a stern propeller may also be used if desired." But now let us turn to the Brightlingsea demonstration.

The *Engineer*'s report observed that "the speed [of the torpedo] can be arranged before it starts [and] after it is once started it cannot be altered." But the speed variations quoted were considerable: "It can be set to run for 300 yards at a rate of 25 knots . . . or for seven miles at 5 miles an hour."[55] Sleeman, however, gives "16 miles [miles per hour?] with a range of 2 miles."[56] For the purposes of comparison, the 1887 Woolwich Royal Laboratory 14-inch Mark VII had a maximum speed of twenty-seven knots but a maximum range of only six hundred yards. The demonstration weapon's spherical engine had been built by Heenan and Froud of Manchester, but according to the report, and despite Lay's claims, the torpedo only reached a top speed of fifteen knots during the trial, although he provided various specious excuses for its poor performance. It was also noted by the writer of the periodical's report that "its most promising function at present, and that in which it best acquits itself, is the guarding of channels," a belated confirmation of Hobart Pasha's verdict some five years earlier.

Sleeman was being very charitable when he described the Brightlingsea demonstration as "not successful." It was, in fact, a farcical failure. The test was supervised and controlled by Lay himself from a small rowing boat, and it was quickly apparent, to even the casual observer, that something was amiss: "The torpedo moved for a short distance and stopped and nothing more could be done at the moment. Moreover the torpedo was not running properly. It was far too high in the water, and far too visible, and it had a list over to one side."[57] Lay explained that the speed of the weapon determined the depth at which it ran, the horizontal rudders having a function similar to a submarine's hydroplanes. Indeed, so far as depth was concerned, the Lay torpedo was no more than a semi-surface weapon which only ran submerged by a combination of speed and angled surfaces rather in the manner of the towed torpedo or the minesweeping paravane both of which relied on the otter-board, a device invented to enable fishermen to keep their trawl nets below the surface. Apparently, a distinguished naval officer who watched the demonstra-

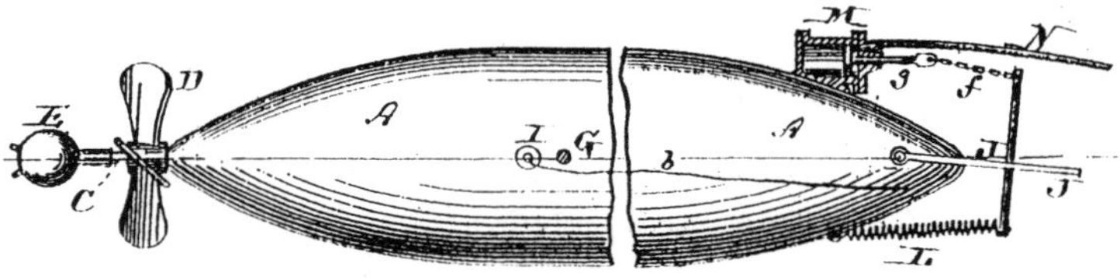

Henry Knapp's float-supported torpedo employed a bow-mounted propeller. Its propellant gas was supplied from an external source via a hose.

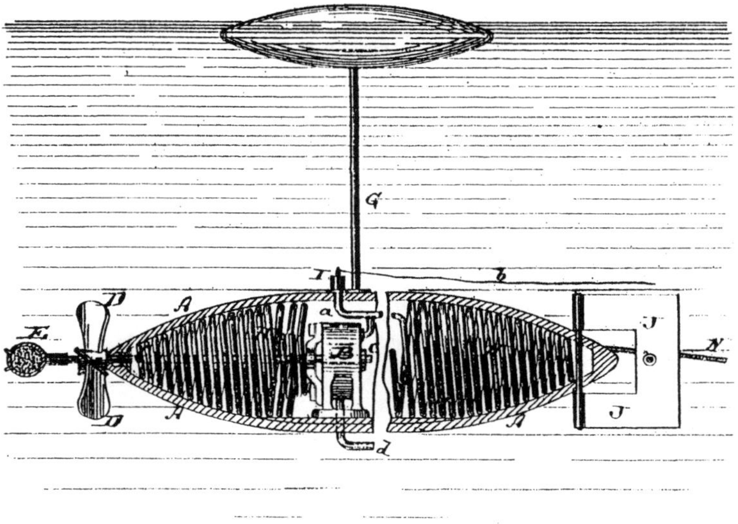

tion was heard to comment that "he would hang any captain who allowed himself to be struck by such a thing."[58]

Lay blamed the mishap on inadequate information that had resulted in the demonstration taking place at low tide, which had led to the screw and cable becoming entangled in weeds and which, in turn, caused a short circuit. The *Engineer*'s observer was sympathetic to Lay and considered the failure to be of small significance. "Awkward management or mistakes of this kind must be distinguished from faults in the torpedo itself. . . . No one would depend on a single torpedo and it seems improbable that had there been more that all would have failed," he wrote. But having piously hoped to see a more successful demonstration of the tor-

Extreme Simplicity · 61

pedo in the near future, the unease of the journalist was apparent in his concluding sentences:

> We say, unhesitatingly, that the behaviour [of the torpedo] must be very different in more than one respect from that of last Saturday's [demonstration]. The torpedo must be very much less visible, and it ought to be started promptly at a given signal, and shown to be well in hand. If at this stage of development Colonel Lay does not exhibit on a show day something that acts promptly and certainly, we shall be inclined to think that he is beset with the infirmity of the late Mr Babbage, who could not be got to perfect one thing because he was constantly tempted by greater future possibilities which some new unworked design presented to his view.[59]

In that final sentence, the writer encapsulates precisely the basic reason for John Lay's continual failure to produce a reliable torpedo. Allied, of course, to his lack of "extreme simplicity."

The technical press and newspapers showed diminishing interest in Lay's activities after the Brightlingsea catastrophe, and it is difficult to find any further mention of either him or his inventions. Having resided in Europe for nearly thirty years, he finally returned home to the United States an impoverished man who, it was said, had lost a fortune "through unlucky speculation." Bearing in mind Victor von Scheliha's experience of Russian chicanery, one wonders whether the unfortunate Lay ever received the full proceeds of his St. Petersburg venture. According to Sleeman, Russia bought no more torpedoes after the initial purchase of ten weapons, and construction of the torpedo factory had ceased well before 1888.[60] From this it can be deduced that the Russian admirals were more than a little unimpressed by the Lay torpedo.

Cruelly aware of his failures, Lay even tried to conceal his true identity in the period immediately preceding his death and was found by an old school friend, Henry Cummings, "lying ill and destitute in a lodging house."[61] He was rushed to Bellevue Hospital, where he died on 17 April 1899.

It was a sad end for a man who had gained his reputation from Cushing's exploits on the Roanoke in 1864. A flawed genius, he proved unable to live up to his early inventive promise.

CHAPTER 4

A Highly Respected Scottish Engineer

THE MYTH that "an employee in the Whitehead manufactory at Woolwich arsenal" had "devised" the system of twin propellers which, working on a common shaft, rotated in opposite directions and improved both the stability and speed of the torpedo was first expounded by Capt. C. W. Sleeman in the second edition of his *Torpedoes and Torpedo Warfare*, published in 1889.[1] And the myth has been accepted and perpetuated by historians for more than a hundred years.

Peter Bethell, in his masterly survey of torpedo development,[2] was deliberately vague about the origin of the concept and chose his words with care: "The later 16-in Woolwich torpedoes had twin contra-rotating propellers, which were embodied by Whitehead in his 14-in model." It was a perfectly sound and unchallengeable statement of fact, although it clearly begged the question of who had been responsible for the innovation. Twenty-five years later, another notable torpedo historian, Geoff Kirby, observed, "With the introduction of the new [Brotherhood] engine and contra-rotating propellers (this latter by a foreman mechanic at Woolwich) no significant improvements were then made until the introduction of the gyroscope . . . in 1895."[3] Unfortunately, he did not reveal the source from which he gleaned the further snippet of information he added in parentheses.

I, too, fell victim to Sleeman's guesswork when I originally wrote *The Devil's Device* in 1974, and it has taken many years to determine precisely who was responsible for this technological breakthrough.[4] In this chapter, therefore, we will trace the origin of the twin contra-rotating propeller concept from John Ericsson's first patent in 1836 to 1874, when

[62]

the Royal Laboratory at Woolwich arsenal tested and adopted the bevel-wheeled contra-rotating screw system devised by Scottish engineer Robert Wilson, an improvement which, with the consent of the British government, was eagerly assimilated by Robert Whitehead himself. Indeed, it proved to be one of the only major advances in torpedo design which owed nothing to either Whitehead or his team of Fiume engineers.

The omission of Robert Wilson, and, similarly, Victor von Scheliha, from previously published accounts of torpedo development seems inexplicable. The work of the latter was clearly known to Cdr. John (Jacky) Fisher, who mentioned his name several times in correspondence, and, of course, his dirigible weapon had been examined by the Admiralty's Torpedo Committee in 1873. In addition, von Scheliha had published *A Treatise on Coast Defence*, which was based on his experiences as chief engineer for the Department for the Gulf of Mexico of the Confederates' army during the Civil War, *and* his torpedo was the subject of a British patent. Wilson's name, too, had be publicly identified in the press when he was granted a monetary award by the British War Department in 1880 for his work on torpedo propulsion, and reference to this official recognition appeared in many of the obituaries published in newspapers and periodicals at the time of his death in 1882.

This is the story of two men. One the renowned Swedish American inventor John Ericsson, the other a leading but little-known Scottish engineer who, at one stage of his career, had been a partner of James Nasmyth.[5] It is a somewhat complicated story, and to understand fully how it unravels, it is necessary to follow the careers of each man so that the validity of the evidence available can be judged one way or the other. By a strange quirk of fate, both men were born in 1803, so, solely on an alphabetical basis, I will deal with Ericsson first.

The son of Olof Ericsson, a mine owner and, in modern terminology, a graduate engineer, John was born on 31 July 1803 at Langbanshyttan in southern Sweden and showed a remarkable aptitude for mechanics at a very early age. Unfortunately, the war with Russia that followed the Treaty of Tilsit seriously affected the Scandinavian mining industry, and by 1811, Olof had suffered financial ruin. He salvaged his fortunes, however, by obtaining employment as an inspector for the Engineering Corps of the Swedish navy, which was, at the time, constructing the Göta Canal,

a daunting project that had been started in 1716 and was not finally completed until 1832, although there were, admittedly, long gaps of inactivity for both political and commercial reasons.

Noting his youngest son's fascination with machinery and his interest in building models and preparing drawings, Olof was able to use his position to obtain cadetships for both John and his elder brother with the Corps of Mechanical Engineers, where he was instructed in algebra, chemistry, geometry, and English. According to one of Ericsson's obituaries, however, he obtained his cadetship through Count Platen, the head of the canal project, who had admired a miniature sawmill and model pumping station which John had built when he was only ten years old.[6] In addition to field training as a military surveyor, he was also taught cartographical and engineering drawing, and, like Robert Whitehead, the beauty of his draftsmanship was highly acclaimed.[7]

Ericsson was barely thirteen years old when he obtained his cadetship, and only four years later he was appointed to a chasseur regiment as an ensign. His prowess as a military surveyor swiftly drew the attention of his superiors, and map making was added to his other duties–this increased burden being alleviated by the doubling of his army pay by putting him on a piecework basis of remuneration–and by 1826, when he left the army, he had risen to the rank of captain. He retained this military prefix for the rest of his life.

Coming to London, he tried to sell his "flame-engine" with markedly little success, for, unfortunately, the device lived up to its name rather too literally. Using coal instead of wood shavings, the fuel for which it was designed, the heat generated was sufficient to melt the engine during a demonstration. And that was the end of that particular project. Next, supported by a partner who financed the venture, he produced a steam fire engine only to have it rejected by the London Fire Brigade, which preferred to put its trust in the old manually operated pumps that had served it so well in the past, provided, of course, that one overlooked the Great Fire of 1666. Despite these setbacks, Ericsson managed to earn his living for the next twelve years as a freelance engineer and inventor of considerable versatility with a track record that was remarkably similar to that of Robert Whitehead's early career in France and Italy between 1846 and 1848.[8] By an extraordinary coincidence, both men found themselves engaged in the design of pumping machinery for draining marshes–Erics-

son in Sweden while he was in the army, and Whitehead in Lombardy, which was, at that time, still part of the Austro-Hungarian empire.

Caught up in the enthusiasm surrounding the new railway age, Ericsson designed and built a steam locomotive, *Novelty* in just seven weeks, and entered it in the 1829 Rainhill locomotive contest. But working to such a tight timetable, and without the track facilities he needed for testing, he was unable to perfect his engine and, even though the locomotive achieved thirty miles per hour, the prize went to George Stephenson's famous *Rocket.* Although in later life the inventor was philosophical about his failure, his biographer was rather more blunt: "With success would have come immediate prosperity and corresponding temptation."[9]

While in England, he designed a rotary engine, developed the use of superheated steam, built tool-cutting machinery, and ventured into many other fields of engineering science. But as time passed, he found himself becoming more and more embroiled in the challenges of steam-power and marine engines. He had already realized that the existing system of steam propulsion, the paddle wheel, was preventing any real advance in ocean navigation and, according to his biographer, he began a series of practical experiments in 1833, when he tested a submerged propeller on the London and Birmingham Canal on behalf of a barge company.

In a letter to John Bourne published some forty years later, Ericsson noted that he had designed a rotary propeller "actuated by steam power" in 1835 and in the following year had constructed a small-scale propeller-driven boat which he demonstrated in a large water cistern preparatory to obtaining letters patent to protect his invention.[10] The patent was not, however, the first screw propeller to be recognized by English law.[11] For just six weeks earlier, on 31 May 1836, Francis Pettit Smith, a farmer from Hendon, Middlesex, had beaten Ericsson to the post with a patent for his "improved" version of the screw propeller–the use of the word "improved" in this context was a legal convention to indicate that Smith was not, in fact, claiming to have originated the concept.[12]

There is, indeed, ample evidence that many people had been experimenting with the idea for a considerable number of years. The Austrians, for example, claim that Joseph Ressel was the first man to make use of the screw propeller principle, as early as 1812, while still a student at Vienna University. In 1826, he began experiments with a barge "driven by hand" and was granted an Austrian patent on 11 February 1827. Two

years later, Ressel fitted his screw to a small boat with a six-horsepower engine and was rewarded with a speed of six knots. Unfortunately, a steam pipe burst and the local gendarmerie, worried about anarchist bomb explosions, which were rife at the time, banned any further experiments.[13] A French army officer, Captain Delisle, also made a written claim that he had devised a submerged propeller in 1823, but this may have been a manufactured story dreamed up by defense lawyers when Ericsson tried to recover damages for patent infringements from France.[14]

The Swedish inventor was, not surprisingly, unhappy about Smith's patent and in 1850 even petitioned the Privy Council to uphold his right to be recognized as the true originator of the screw.[15] He was supported by John Bourne, the author of *A Treatise on the Screw Propeller* and an expert on the history of screw propulsion, who wrote, "Ericsson . . . was an accomplished engineer; Smith was only an amateur . . . [but] Ericsson's mechanical resources gave him [the] means of overcoming difficulties such as Smith did not possess."[16]

In an earlier affidavit, sworn in March 1845, Ericsson produced irrefutable evidence that he had demonstrated his screw propeller on the London and Birmingham Canal on a number of occasions *before* 1833.[17] And he named Elias Harrison, subsequently a chief engineer in the U.S. Navy, as the man he employed to fit the screws to the canal boat *Francis,* which acted as the test bed for the experiments. Finally, he supplied the names and addresses of the principals for whom he was carrying out the tests— Robins and Mills of London Wall in the City of London. Smith's patent, however, remained undisturbed, despite this legal furor about its validity.

It may seem odd to examine the invention of the screw propeller in a book primarily concerned with the historical development of the torpedo, but it is, in fact, relevant, for Ericsson's 1836 patent was for *contra-rotating* propellers, not for screw propulsion per se. In addition, the other main protagonist in the controversy about the adoption of contra-rotating propellers for both the Woolwich and Whitehead torpedoes, Robert Wilson, also proclaimed himself to be the original inventor of the screw propeller and in 1880 produced a sixty-page monograph, which he extended to eighty pages in its second edition, in support of his claim.[18] Wilson's work will be examined later in this chapter. At this point, however, it can be revealed that a letter which Wilson reproduces in his monograph confirms beyond question that he was the person responsible for

the Royal Laboratory testing, and then adopting, twin contra-rotating propellers for its 16-inch torpedo.

Apart from the greater efficiency and improved performance of the screw propulsion system, Ericsson had already recognized the military advantages that would follow from its adoption. Initially the Royal Navy's admirals had opposed the use of steam power in warships on the not unreasonable grounds that the large paddle-wheel boxes and their associated deck-mounted machinery would be an easy target for enemy guns, and for that reason alone, experienced seamen continued to favor sails. Steam power's reliance on fuel, with its consequent limitation of range, also did not pass unnoticed. The Swedish inventor realized, however, that, while he could do nothing about the reliance on fuel as opposed to the natural element of the wind, the use of a submerged screw meant that the engines could be placed below the waterline, where they would be totally protected from the guns employed at sea at that particular moment in time. This meant that the admirals had one less plank on which to rest their opposition to steam warships.

Ericsson is, of course, most famous for designing the *Monitor,* the world's first turret ship whose guns enjoyed a 360-degree uninterrupted arc of fire—a revolution in design as important as the transition from sails to steam power. In fact, it proved to be the progenitor of the modern warship. The turret guns also provided another incentive to do away with the impediment of sails and adopt the steam engine with its concomitant submerged screw propeller which left the decks relatively clear of clutter. Ironically, as ironclad design developed, the desired 360-degree arc of fire was never again achieved.

Contrary to popular belief, the turret was not invented by Ericsson. The revolving gunhouse employed on the *Monitor* was initially designed by Theodore R. Timby of New York City in 1841, and he obtained a patent for his invention in 1843, after the U.S. War Department had rejected the idea. An improved version, described as a "revolving battery tower, was patented in 1862. In truth, this so-called improvement was something of a monstrosity with its numerous guns, mounted in tiers like an ornate wedding cake with the cannons set around its central axis as if they were the spokes of a wheel.[19] As with so many other nineteenth-century inventors, Timby's attempts to make his device better merely ended up with it disintegrating into impractical complexity, rather like von Scheliha's

efforts to transform his perfectly workable torpedo into a multiheaded war machine intended to destroy a complete fleet all on its own.

Ericsson wisely came down on the side of simplicity and opted for just two gun barrels in his turret. But Timby promptly claimed the rights to the "revolving tower" because one of his drawings had shown it being used at sea on a low-freeboard ship very similar in appearance to the *Monitor.* Ericsson did not bother to argue—time was too precious for prolonged litigation. And, possibly, Timby had overlooked that there was a war going on. So settling out of court, he paid the inventor an agreed royalty on each and every turret he built.

The ultimate defeat of the Confederate *Virginia* (the former Union *Merrimack*) during their historic battle in Hampton Roads on 9 March 1862 spelled the end of the broadside ironclad, though many years were to pass before its final demise. Indeed, secondary armament guns were still mounted on the beam of major warships throughout and beyond World War II, which meant that, in a surface action, only half of those particular guns could be employed if the vessel maintained her place in the line of battle. Nevertheless, even if he had done nothing else, the *Monitor* had secured John Ericsson's place in naval history.

The successful application of his screw propeller to ships, especially those engaged on the North Atlantic routes, falls outside the remit of this particular work but, again, added further to his reputation as one of the great inventors of the nineteenth century. He died in New York on 8 March 1889, and although seriously ill with a degenerative kidney disease, continued to work on his latest project, a solar-powered motor. His body was returned to his native Sweden for burial.

Ericsson's pneumatic torpedo was examined in chapter 1, but it is necessary, at this point in the narrative, to turn the clock back even further and to consider the details of the 1836 patent for his first screw propeller. After the usual introduction, it continues: "I, the said John Ericsson, do hereby declare the nature of my said invention to consist of two thin broad hoops, or short cylinders, made to revolve in contrary directions round a common center, each cylinder of hoop moving at a different velocity from the other, such hoops or cylinders being also situated entirely under the water at the stern of the boat, and furnished each with a series of short spiral planes or plates, the plates of each series standing at an angle the exact converse of the angle given to those of the other series."[20]

The remainder of the patent is taken up with a technical description of the drawings that formed an integral part of the specification. It is interesting to note two things. First, the concept of twin contra-rotating propellers was part and parcel of Ericsson's first screw propeller system and not, as one would expect, a subsequent development from an original single screw propeller design. Second, Ericsson does not make any specific claims regarding the advantages to be obtained by the screw system over that of the existing paddle-wheel method of propulsion other than the brief comment that "a steamboat may be propelled effectually, notwithstanding any variations in its draught of water."

The drawings with the patent specification show that contra-rotation of the two screws was obtained by means of two straight cogs which transferred the rotation of the lower shaft to the hollow shaft working one propeller, while its twin was driven by an inner lower shaft running directly from the engine. This mechanism is relevant to the torpedo, for, unmodified, it would have probably been impossible to fit it inside the steel shell of a standard 16-inch fish torpedo of the early 1870s. And as will be seen later, it was Wilson's improvement that made the adoption of twin contra-rotating propellers for the torpedo possible.

Considerable efforts have been made to find either an interior drawing of Ericsson's pneumatic weapon or a subsequent patent modifying the 1836 straight cog system, but to date the search has yielded no results. As already noted in chapter 1, the pneumatic torpedo proved to be impractical and virtually unmanageable in the water due to the "drag" effect of the external air hose. Its performance, too, was unacceptably poor; Sleeman gave its range as only eight hundred feet with a maximum speed of three to four knots.[21]

If Ericsson failed, or neglected, to miniaturize his cogwheel system by adopting bevel gears, as Wilson was to do with the Woolwich torpedo, this could explain the ungainly proportions of the pneumatic weapon, which was more box-shaped than cylindrical, unlike the streamlined Whitehead torpedo. While considering this aspect of the contra-rotating propeller mechanism, it is interesting to note that even Francis Pettit Smith, despite his scorned amateur status, incorporated bevel gearing into the shafting of his single-screw propeller, according to the drawing appended to his patent.[22]

But Ericsson was already losing interest in propellers by the time his

pneumatic torpedo came before the public, and his mercurial mind had moved on to the concept of the projectile weapon which will be considered in more detail at a later stage of this chronology.

Robert Wilson's exact date of birth is not recorded, but he is known to have been born in the same year, 1803, as John Ericsson at Dunbar in Scotland. He was orphaned while still a child when his father, a local fisherman and volunteer lifeboatman, was lost in the course of a heroic attempt to reach the stranded frigate *Pallas* after it had run aground on the East Lothian coast in December 1810. Two successful trips had been made in heavy seas to take off the crew but on the third attempt to rescue the remaining survivors tragedy struck and Wilson's father was swept out of the boat and drowned.[23]

His first thoughts on screw propulsion, if we are to believe his account, came to him when he was only five years old. "The idea of the screw propeller took possession of my mind as early as 1808," he wrote.[24] A soldier stationed in Dunbar had fitted rudimentary paddle wheels to a small fishing dinghy and Wilson noticed that, although effective in smooth water, they were not suited to a seaway. And, an accomplished sculler himself, despite his extreme youth, he pondered whether something similar to a sculling oar could be fitted to the stern of the boat.

In spite of his subsequent professional achievements in later life, Wilson received only a "meagre education" and soon after his father's death left Dunbar to be apprenticed to a joiner.[25] According to his own account, the problems associated with marine propulsion continued to occupy his mind. One day, in 1812, while out for a walk, he encountered a running stream acting upon an undershot waterwheel, "a piece of machinery I had never seen before."[26] He recognized a passing similarity with the paddle wheels fitted to the soldier's boat but quickly appreciated that, in the case of the mill, the water acted upon the wheel rather than vice versa.

"Shortly afterwards . . . I saw, on the farm of Oxwellmains, a windmill used for threshing corn," he confided. "And on inquiry I learnt that it reefed and unreefed its own sails, and turned its face always towards the wind–*all by self-action or mechanical control.*"[27] He returned a few days later with a telescope and carefully examined the mechanical arrangement of the windmill, from which he deduced how to modify the sculling

oar into the form of a windmill's sails "with its actions reversed." Bearing in mind that Wilson was barely nine years old at the time, his mature and almost inspirational reasoning, suggests extreme precocity.

Due to lack of funds and the necessity of earning a living, Wilson contented himself by making a small model with a set of four blades in wood. And there matters rested until 1821, when a Scottish paddle steamer got into difficulties off Dunbar as a result of its paddles lifting out of the water when the vessel rolled excessively in a groundswell. Wilson personally witnessed this frightening incident and realized that his submerged screw propeller would have averted the danger in which the unfortunate *Tourist* now found herself. And the experience prompted him to return to his experiments with what he termed "Rough Sea or Storm Paddles." He proceeded to build several more test models, but, again, the demands of earning a living forced him to set his work to one side after a few months.

It would be tedious to follow in painstaking detail Wilson's account of his progress in producing a practical screw propeller. The full story is given in his monograph, or, as he preferred to describe it, pamphlet.[28] But the following outline will examine the evidence he produced to support his claim to have invented the device before either John Ericsson or Francis Smith.

Between 1825 and 1827, he built and experimented with various versions of the propeller in the course of which he found that a two-bladed screw was more mechanically efficient than the three or four blades he had employed hitherto, and that by placing the propeller "behind the rudder,"[29] he achieved improved performance, a conclusion shared by Robert Whitehead.[30] The Royal Laboratory at Woolwich, however, disagreed and preferred the opposite configuration of the propeller abaft the rudder.[31] By this time Wilson's models were being powered by clockwork and this enabled him to observe the action of his screw propellers with considerable accuracy.

The looseness of the terminology used with regard to the positioning of the propeller vis-à-vis the rudder can be extremely confusing, for it depends on how you view the configuration whether the propeller is in front or behind the rudder. It is easy to grasp when confronted with a diagram but ambiguous textually without firm guidance regarding the viewpoint adopted. If the bow of the vessel or torpedo is pointing to the

72 · *Nineteenth-Century Torpedoes and Their Inventors*

right, the sequence at the stern is (1) propeller to the right of the rudder, that is, it is nearer to the ship than the rudder, or (2) propeller to the left of the rudder, with the latter being closest to the ship.

It will become instantly apparent that (1) could be described as a propeller *behind* the rudder, and yet, if viewed relative to the rudder, it could equally be correctly described as *in front* of the rudder. The configuration adopted by both Wilson and Whitehead was (1) while the option employed by the Royal Laboratory was (2).

At this stage of his narrative, Wilson produced precise and firsthand evidence in support of his claims. In 1827, he was introduced to James Hunter, president of the Dunbar Mechanic's Institution, who in turn secured him an audience with the earl of Lauderdale. The nobleman showed great interest in Wilson's ideas and believed the young engineer deserved encouragement. But before using his influence with the Admiralty, he asked the Honorable Antony Maitland, at that time the commanding officer of the frigate *Glasgow,* to observe a test of the screw on a local "sheet of water." The experiment was a great success, and the various members of the institute who were present "expressed themselves as delighted the results."[32]

Maitland, supported by Sir William Houston, reported to his father accordingly, and Wilson was invited to the Earl's residence the very next morning, where he was informed that His Lordship "was convinced the invention was worthy investigation."[33]

Yet to Wilson's dismay, the Admiralty showed no interest and even declined to witness a demonstration of his invention. "After this," he wrote, "I determined to make my invention publicly known, for although many persons had witnessed the experiments with the model, as yet no public notice had been taken of them."[34] On 18 October, Wilson therefore took steps to have a record entered into the minute book of the Dunbar Mechanic's Institute. Because of its importance in substantiating the inventor's claims, it is reproduced here in full:

Mr Robert Watson presented to the meeting two models invented by Mr Robert Wilson (member of this Institution)—the first, a Horizontal Wind Wheel; the second, a Model with an Apparatus for propelling Steam vessels from the stern, a kind of Revolving Scull. By it the vessel goes with greater speed than with side paddles, and produces so little motion of the

water as to fit admirably for canal navigation. About three months ago, this model, with the apparatus attached, was tried on the water in the presence of the Honourable Capt. Antony Maitland and Sir William Houston, as well as several leading members of this institution.

(Signed) William Robertson, Secretary
[Entered 18 October 1827][35]

To Wilson's undoubted satisfaction, news of the invention also reached the press, and on 29 December 1827, the *Edinburgh Mercury* carried the story and added, "We have since had the satisfaction of examining the model . . . [and] we consider it as completely adapted to the objects of the ingenious inventor–particularly that of giving to canals the advantage of steam navigation."[36] This latter benefit is often overlooked when the usefulness of the screw propeller is being assessed. Despite the rapid growth of railway freight traffic, inland waterways still played an important role in the nation's economic infrastructure and the erosive effect of paddle-wheel propulsion effectively debarred steam power from Britain's canal system.

In the following year, Wilson obtained the sponsorship of the Highland Society of Scotland, which "allowed the sum of £10 towards enabling Mr Wilson to get paddles or 'propellers' made on a large scale, with corresponding machinery, [and] to be applied to a boat."[37] This twenty-five-foot-long craft was powered by a two-man team working the hand crank that rotated the propeller shaft and the demonstration, which took place off Leith, was observed by a number of important witnesses, including Vice Adm. Sir David Milne, in April 1828. The tests were successful, or, as the society's secretary put it in his report, "The results . . . answered our most sanguine expectations," and the society agreed to pay Wilson a further sum of ten pounds "to defray the expenses of the experiments," although this was conditional upon him allowing them to acquire "my model."[38] The inventor was reluctant to sell the model on which he had lavished much "anxiety and labour." But as he readily admitted in his monograph, "I had, however, no alternative between doing this or what I dreaded still more, getting into debt."[39]

Despite the success of the Leith demonstration, the impecunious inventor could find no one willing to either finance or build a full-sized vessel to exploit the obvious advantages of his screw propeller. Once

again, Wilson was forced to set his invention to one side and return to the more mundane task of building up his business as an engineer.

Thanks to the efforts of his friend James Hunter, interest in the screw propeller was revived in 1832 when it was brought to the attention of the Society of Arts for Scotland, which agreed to conduct further experiments for which it would defray the costs. Interestingly, the propellers were still referred to as "Stern Paddles" rather than screws. The tests, using a loaned eighteen-foot boat, were held on 7 and 18 June 1832 and deemed so successful that the society rewarded Wilson with a silver medal (valued at five sovereigns) which was presented to him at its annual general meeting on 12 December. Wilson even included a sketch of the suitably inscribed medal in his monograph.

Realizing that it was essential to carry out further tests in a much larger vessel, and appreciating that this could only be done with the aid of government finance or facilities, the society took its case to the highest levels only to have it rejected by the Duke of Richmond. "I regret that it is not within my power to comply with your request," he wrote Sir John Sinclair, who was acting as an adviser to the society. "The regulations of the department do not permit me to make experiments in the packets for the obvious reason that, if the experiment should fail, the consequent delay would be a serious inconvenience to the public."[40] This rather strange excuse was presumably in response to a suggestion that an experimental screw propeller should be fitted to a government-owned Post Office mail boat. Richmond was postmaster general in Earl Grey's government from 1830 to 1834 and thus responsible for the mail service. This was, of course, before the introduction of the Penny Post in 1840.

An approach to the Admiralty proved equally disappointing. The propellers, machinery, and documentation from the Leith demonstration were referred to Woolrich dockyard and on 17 September 1833, the captain superintendent replied, "We have carefully examined the papers . . . [and in our] opinion . . . the plan proposed . . . is objectionable, as it involves a greater loss of power than the common mode of applying the [paddle]wheels to the side, and herewith return the papers."[41]

Wilson added a footnote that "neither the propellers nor the machinery were returned with the papers here referred to. It would be interesting to know what became of them, so as to judge of the opportunity afforded to others to improve upon them."[42] His suspicion that his screw propeller

became the basis for Smith's subsequently successful patent is clearly apparent and barely concealed, though there is no real evidence to support the allegation.

It was, nevertheless, an extraordinary conclusion from what, after all, were the British government's experts. Yet no attempt was made to justify the dismissive claims regarding "loss of power" either by way of practical demonstration or theoretical calculation. For Wilson it was the end of his hopes. "The unfavourable reception of the invention met with from the Admiralty . . . deprived me of the cooperation of the noblemen and gentlemen who had until then actively countenanced me. Seeing no prospect of bringing the invention to a practical development . . . they gave up the subject in despair, expressing great regret and disappointment that their influence and trouble had been spent in vain."[43]

A weaker character could have been broken by this final turn of events, but Wilson was made of sterner stuff and, hiding his anger at the Admiralty's rebuff, he "determined . . . to wait with patience for an opportunity of getting some shipbuilder or other enterprising person to take up the matter."[44]

Fate was to deal him an even greater blow three years later, when the little-known Francis Pettit Smith was granted the world's first patent for a screw propeller. As already noted, this was followed only six weeks later by John Ericsson's patent for twin contra-rotating propellers. In hindsight it was clearly foolish of Wilson not to have applied for patent protection as soon as he had proved that his propeller worked effectively. He later admitted the error but explained that his failure to apply for letters patent was simply due to lack of the money necessary to cover the legal and other fees involved. In fact, he did not take out his first screw propeller patent until 1861.[45] But it was now too late and the glory, not to mention the financial rewards, went to Pettit Smith, who was later honored with a knighthood.

The drawings accompanying Smith's patent (No. 7104 of 1836) showed it to be a true screw. The technical description of the specification reads, "A screw of one turn or two half turns." Wilson's device was, of course, in the form of two scull-shaped blades. It is also interesting to note that Smith's mechanism incorporated two sets of bevel gears while Wilson, like Ericsson, employed straight cogs. The significance of the gear wheels will become apparent when we examine the twin contra-rotating pro-

peller system which was ultimately adopted for the Woolwich and White-head torpedoes.

Before closing this section on the origins of the screw propeller, a few observations about Pettit Smith, as he was usually known, will be of value in understanding both Wilson's and Ericsson's angry reaction to his patent. A grazing farmer from Hendon in the county of Middlesex, Smith was twenty-six years old when he first became interested in screw propulsion. Coincidentally, he was born in 1808, the year in which Wilson first realized the advantages of "stern sculls." Within two years he had solved the problem, and by the spring of 1836, he had built a model boat, fitted with his screw, which was tested on a pond at his farm and exhibited at the Adelaide Gallery in London. With the help of a Mr. Wright, "a banker," he applied for and obtained a patent for his "improved propeller" in May of that year. Now that Smith was backed by "rich and enterprising commercial men" and aided by what would now be called a public relations campaign, the Admiralty, by 1838, found itself committed to the construction of screw ships.[46] According to Wilson, who supported his allegations with comparative drawings,[47] Smith had, by 1845, modified his original patented screw into a propeller identical to that used by the Scottish inventor at the Leith trials in 1827. Wilson's anger and suspicions are, to say the least, understandable.

Shortly before this, in 1843, the 888-ton sloop *Rattler* was fitted with Smith's propeller and a tug of war was arranged by the Admiralty with the paddle-wheeled *Alecto,* a contest which *Rattler* won hands down when the screw-driven sloop dragged its hapless rival stern first at two and a half knots in a demonstration as dramatic as any witnessed before and which proved, beyond argument, the superiority of the screw propeller over the old side-mounted paddle wheels.

In his monograph, Wilson hints at some behind-the-scenes skulduggery which led to Pettit Smith winning the Admiralty contract. Apart from his own experiences, he instanced John Ericsson's similar lack of success. For although the Swedish engineer had obtained his patent only six weeks after Smith and had demonstrated the efficiency of his invention by towing a barge with the lord commissioners of the Admiralty aboard up the Thames at a rate of ten knots behind his first experimental steamboat, the *Francis B. Ogden,* in 1837, he received no official encourage-

ment.[48] Giving up in disgust, Ericsson finally made his way to the United States as soon as his first full-sized screw-propelled vessel, the *Robert F Stockton*, was completed in 1839.[49] Britain's loss was undoubtedly America's gain.

Before moving on to the use of contra-rotating propellers in torpedoes and, in particular, resolving the question of who actually proposed and designed those adopted by the Royal Laboratory at Woolwich in November 1874, a very brief summary of the situation vis-à-vis the screw propeller will not come amiss.

Based upon the evidence so far adduced in this chapter, it seems reasonable to submit that Robert Wilson was the true inventor of the modern screw propeller but Francis Pettit Smith was accorded the honor because he was the first to patent the idea. Furthermore, Ericsson's 1836 patent proves beyond question that he was the originator of the contra-rotating propeller concept. The importance of establishing these facts will become apparent shortly.

Undismayed by his lack of fortune with the screw propeller, Robert Wilson continued to enlarge his reputation as an innovative mechanical engineer. He moved from Edinburgh to Manchester, where he was appointed manager of the Bridgewater Foundry, owned by the eminent engineer and inventor James Nasmyth. According to the *Dictionary of National Biography,* he assisted the latter to perfect his famous steam hammer by designing its self-acting motion. And after a spell at the Low Moor Ironworks near Bradford, during which time he added the "circular balanced valve" to the steam hammer, he became the managing partner in Nasmyth, Wilson and Company when Nasmyth retired in 1856, a clear indication of his esteem in the world of engineering.[50]

The *Dictionary of National Biography* entry also noted, "He afterwards constructed the great double-acting hammer at Woolwich Royal Arsenal . . . [and] in 1880 the War Department made him a grant of £500 for the use of his double-action [*sic*] screw propeller as applied to the fish torpedo." It was this sentence, and the other obituaries published after his death in 1882, which at first suggested that Wilson might have been "one of the employees in the Whitehead manufactory at the Woolwich Arsenal" to whom Sleeman referred in his book. As he had worked at Woolwich during the installation of the Nasmyth steam hammer, Sleeman's

error could be regarded as understandable, but the path of Wilson's career made it clear that he had never been an employee of the Royal Laboratory. And the discovery of a patent which Wilson had been granted on 23 June 1876 only served to deepen the mystery of the government grant he was awarded in 1880.[51]

This patent was concerned with "improvements" to propellers, in particular, overcoming the loss of speed occasioned by the "slip" of the screw. He added significantly, "This invention is also particularly applicable for propelling the fish torpedo, as by it the highest possible speed can be obtained." However, the patent did not result in positively identifying Wilson as the man responsible for introducing contra-rotating propellers to the Woolwich torpedo for two reasons. First, the Royal Laboratory, as we will see, tested the twin propeller system in 1874, and it was already in production by 1876. Second, and more important, the system applied to the RL 14-inch Mark I weapons incorporated *three* miter wheels[52] while Wilson's patent was based upon *four* miter wheels, which meant that someone had designed a contra-rotating propeller system before Wilson's patent was obtained.

Nevertheless, the 1876 patent, and Wilson's comments in his monograph, throw a new–and dare one say, dubious–light on his character, and this cannot be passed unremarked in view of later developments in the contra-rotating propeller story. Wilson was, of course, aware of Ericsson's contribution to screw propulsion and, indeed, held him in high regard for in his monograph he refers to the Swedish engineer as "this accomplished mechanician" as well as mentioning his 1836 patent.[53] Furthermore, in a footnote, he compares Pettit Smith's demonstration boat to "the more efficient vessel of Ericsson's."[54]

In view of Wilson's obvious respect for Ericsson, it thus seems a little strange that he neglected to mention that the Swedish engineer's original screw propeller was of a contra-rotating type or that his 1836 patent was incontrovertible evidence that the concept of two propellers working on the same shaft yet revolving in opposite directions originated with Ericsson as well. To make matters worse, the wording of Wilson's 1876 patent makes it clear that the credit for discovering and adopting the principle of twin contra-rotating propellers was his and his alone: "Now my invention consists in so arranging, constructing, and applying screw propellers in certain positions and in relation to each other, and in so actuating

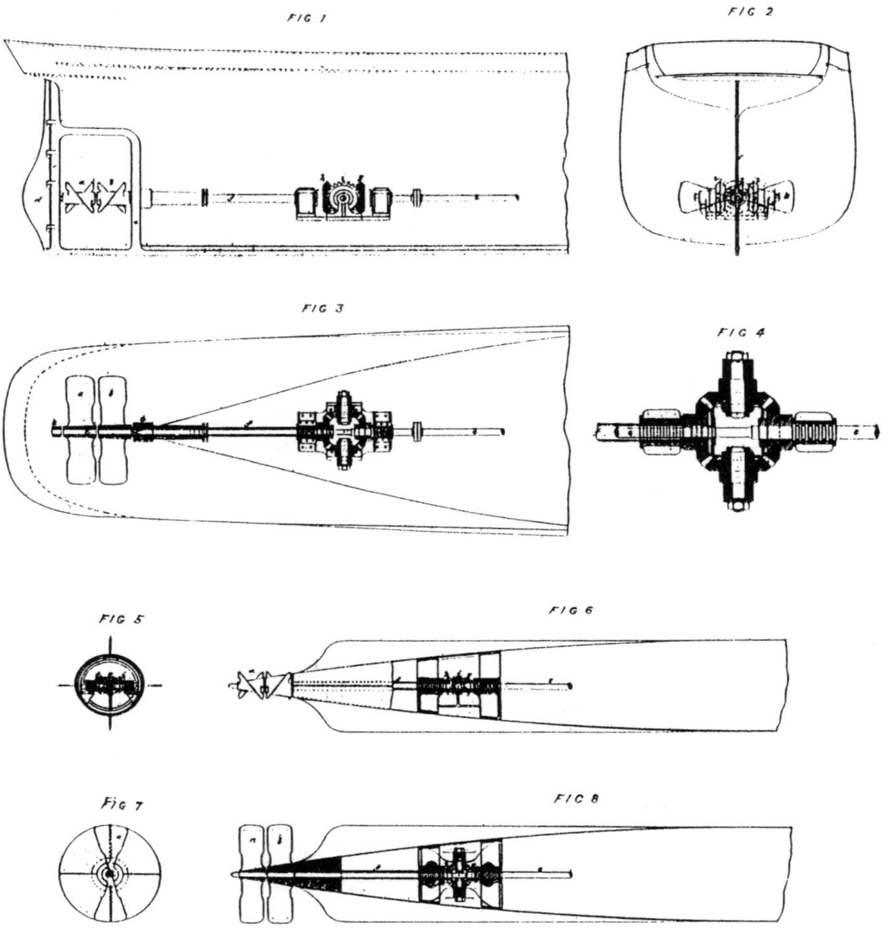

Robert Wilson's improved twin contra-rotating propeller systems, based on four-miter wheels and first used in the 14-inch Mark II RL weapon. His original mechanism employed three-miter wheels.

them that the slip is reduced to the smallest possible amount." The only possible conclusion that can be drawn from this statement is that Wilson's frustration at being denied credit for inventing the screw propeller had led him, probably unintentionally, to steal another man's thunder. Certainly Wilson's 1880 monograph gives the impression of a person almost mentally unhinged by the unfairness of both his treatment by the

80 · *Nineteenth-Century Torpedoes and Their Inventors*

Establishment and the inequalities of the patents system as it existed in the middle of the nineteenth century.

The slowly accumulating body of evidence was, nevertheless, beginning to point to Robert Wilson as the man responsible for introducing contra-rotating propellers to the Woolwich torpedo. And this conclusion seemed to be supported by the British government's monetary award to him "for the use of his double-action screw propeller as applied to the fish torpedo" in 1880.

But a firm attribution remained uncertain, for, as already noted, Wilson's 1876 patent related to "improvements" to the torpedo's existing propulsive system and introduced a four-bevel wheel mechanism to supersede the current three-bevel wheel arrangement. The date of the award (1880) also suggested that it related to the 1876 patent rather than the introduction of the first contra-rotating propulsive system in 1874.

The discovery of a letter written by John Ericsson on 31 July 1878 similarly appeared to further distance Wilson from involvement in the initial adoption of twin propellers by the Royal Laboratory: "The most important part of the whole contrivance is my instrument for driving the fish ahead. You are aware that *since Mr Davidson of Woolwich dockyard copied my double torpedo propeller* Whitehead has also adopted the plan of employing two propellers revolving in opposite directions round a common center, one behind the other."[55]

This latter statement was accurate, for among papers found at the Public Record Office was a letter from Robert Whitehead to the Admiralty dated 11 September 1876, part of which read, "Having heard of the excellent results as to speed and lateral direction obtained by the adoption of the two screw propellers, working in opposite directions, I wish to try this system on my torpedoes and if found superior, I would, with their Lordship's permission introduce it into all the torpedoes which their Lordships have honored me with the supply."[56] The order to which Whitehead was referring related to the construction of two hundred Fiume 14-inch Model A (classified by the Royal Navy as Fiume 14-inch Mark I) torpedoes, delivery of which commenced on 26 May 1877.[57]

There was a flurry of internal minutes, many now unfortunately illegible, on receipt of this letter at the Admiralty. Some officers were reluctant to grant permission while others, aware that two hundred weapons were on order from the Fiume factory, appreciated the benefit to the Royal

Navy if they were equipped with the new twin propellers before delivery. There was, however, another factor in the equation. At this particular time, the Woolwich version of the Fiume torpedo was suffering from "vertical deviation"—or erratic depth control—and as Whitehead had apparently solved the problem, it had been suggested that he should be offered "a small sum of money" for revealing his solution.[58] There were concerns, too, about the terms agreed for the two hundred weapons already on order. Whitehead had proposed a minimum price of £350 per torpedo but had reserved the right to increase this to £387 and 10 shillings if necessary—an increase in total of £7,750, which would have to be met from the public purse.

Unfortunately, the subsequent correspondence cannot be traced, but Whitehead was finally granted the requisite permission, probably as a quid pro quo for exchanging the secrets for controlling the problem of vertical deviation. And also possibly for agreeing in return to stick to his minimum price option for the two hundred–strong delivery that was to follow.

But to return to the accusation that Ericsson had made in the previous part of his letter. Who was the mysterious "Mr Davidson of Woolwich dockyard," whom, he alleged, had purloined his contra-rotating propellers? A search of documents at the British Patent Office revealed a James Davidson "of Woolwich" who had designed and patented several pieces of machinery for the manufacture of center-fire cartridges and this certainly made the Woolwich dockyard and Royal Arsenal connection indicated in Ericsson's letter a strong possibility. Could it be that James Davidson was Sleeman's mythical "employee in the Whitehead manufactory at Woolwich arsenal" and, if he was, then Robert Wilson was not the man responsible for the adoption of contra-rotating propellers by the Woolwich and Fiume factories.

A small news item in the *Times* of London relating to a demonstration of the first Woolwich-built Whitehead torpedo to be completed by the Royal Laboratory, which took place on 28 March 1873 for the benefit of George Joachim Goschen (later Viscount), first lord of the Admiralty, and other distinguished personages provided details of his identity: "Mr Davidson, manager of the department [the Royal Laboratory]; Mr Lowe, one of the foremen whose mechanical skill is said to have contributed greatly to the efficiency of the torpedo; and a number of workmen to assist were

82 · Nineteenth-Century Torpedoes and Their Inventors

[the only other people] present."[59] By coincidence, an engraving reproduced in *The Devil's Device* actually depicts the demonstration on the military canal at Woolwich arsenal and there can be little doubt that both James Davidson and Mr. Lowe appear in the group of men clustered around the torpedo. The date cited in the caption, May 1872, is clearly incorrect in the light of subsequent research and probably relates to its publication date in the *Illustrated London News*.[60]

This seemingly innocuous newspaper report is, nevertheless, of considerable historical importance. First, it indicates that the object of Ericsson's angry accusation; "Mr [James] Davidson" was, in fact, the manager of the Royal Laboratory. It is also apparent that the Swedish inventor had no real idea who was responsible for "borrowing" his ideas and personally blamed the unfortunate Davidson simply because he happened to be the manager of the Royal Laboratory at the time the propellers were adopted.

Second, it identifies the foreman, Mr. Lowe, as being responsible for the modifications carried out on the Whitehead torpedo at Woolwich before manufacturing actually started—most important, the alteration to the propeller and rudder configuration to which reference was made earlier. The date of the report and the demonstration also makes it clear that Lowe's improvements could not have included the adoption of contra-rotating propellers as these were not tested, as will be seen shortly, until November 1874. That Lowe's work was clearly well known to the torpedo fraternity at that time is apparent from the *Times*'s report, and this no doubt explains Sleeman's uncharacteristic error in attributing the adoption of the twin propellers to an employee of the torpedo factory at Woolwich arsenal.

A letter, dated 23 June 1877, from Col. G. Fraser of the Royal Laboratory to Robert Wilson, provides irrefutable confirmation that the contra-rotating propeller system adopted at Woolwich was designed and submitted by Wilson in 1874. Because of its importance to the story of torpedo development, the letter is reproduced in full:

23 June 1877

Dear Sir,

In reply to your letter of 19th instant, I beg to inform you that in November 1874 your double screw was put on a torpedo which, with the single screw

in use before that date, had given a speed of 10.24 knots for 200 yards at 30 atmospheres; the result was that a speed of, 12.33 knots was obtained for the same distance and at the same pressure. The average speed of torpedoes of this class since made by this department has been 12.5 knots.

And believe me,

Yours faithfully,

G. Fraser Col.[61]

The last pieces in the jigsaw puzzle were put into place with the chance discovery that although the original system as fitted to the 14-inch RL Mark I–and the earlier, obsolete 16-inch model–was based on a three miter wheel gearing, the four miter wheel arrangement which Wilson had patented on 23 June 1876 was fitted to the RL 14-inch Mark II weapon, which came into service in 1879 and 1880. This had not been noticed earlier, for the *1887 Torpedo Manual* only illustrated the propeller shafting of the Mark I and omitted the Mark II, which led to the mistaken assumption that the two were identical. The relevant section on the Mark II reads, "The propeller gearing consists of four mitre wheels; the two intermediate wheels are placed one on each side, and held in position by a cross-head which fits into two projections on the after end of the buoyancy chamber, and which also acts as a distance piece."[62]

And so Sleeman was wrong. It was not a Woolwich mechanic who was responsible for the Woolwich torpedo's contra-rotating propellers but a highly respected Scottish engineer whose other achievements had been acknowledged and honored by his peers.

Although not wishing to detract from his reputation, it was a pity that Robert Wilson failed to mention the fact that John Ericsson had fitted contra-rotating propellers to his pneumatic torpedo at least five years earlier. But as we have seen, he had a boulder-sized chip on his shoulder, and any credit, even if wrongly asserted, was better than none.

It is a matter of historical fact, however, that various manually operated propellers were employed during the eighteenth century, notably in Bushnell's submersible *Turtle* of 1775, nearly thirty years before Wilson was born, and Robert Fulton's *Nautilus* in 1798. So although Wilson was responsible for the introduction of twin contra-rotating propellers to the Woolwich-Whitehead torpedo, and despite his assertions to the contrary, he was most certainly not the inventor of the screw propeller. Neither, for

that matter, was Ericsson or Pettit Smith. The credit probably belongs to David Bushnell, although it is equally possible that the manually operated propeller was already in existence when he adopted it for use in *Turtle*. At the time of this writing, its true origins remain unrecorded and unknown.

CHAPTER 5

Most Splendid Results

COL. VICTOR VON SCHELIHA's torpedo and John Lay's movable torpedo submarine, despite their ultimate failure as practical weapons, opened the floodgates to inventive mayhem and, over the next two decades, hard-bitten engineers, armament kings, and optimistic dreamers–not forgetting a handful of complete madmen–tried to join the race. It was common knowledge that Robert Whitehead had made a fortune in double- quick time. And as the world's navies fell over themselves to acquire even larger arsenals of this latest weapon, there seemed more than enough honey in the pot for everyone. The result was a plethora of patents, plans, devices, and unfulfilled promises. The following chapters will seek to create some form of order from the chaos of nineteenth-century torpedo development.

The first subject on the list must be the Punshon, which received a passing mention by A. L. Alliman in his King's College Engineering Society's lectures–the same series of talks noted in chapter 2 in which von Scheliha's name also made an early appearance. In view of the misspelling of the latter, it is probable that Punshon was a misspelling of Paulson, of whom more will be learned later, when we reach the 1880s. In support of this conclusion, it can only be observed that not a single trace of Punshon has been found in twenty years of research. The third name quoted by Alliman was Cave. But, again, it has proved impossible to identify either this gentleman or his torpedo. Misspellings, which seem common in this period, can lead to much unproductive work, so it is as well to dispose of another example before proceeding further.

Bethell describes the Borden in part 5 of his series of articles on torpedo development, although he puts no date to it.[1] However, a careful examination of the specification he provides suggests that this was, in fact,

[85]

the Berdan, as many of the features, particularly the rocket-turbine propulsive system, are common to both weapons. Also, there is independent corroborative evidence of the Berdan torpedo's existence by way of contemporary newspaper reports and the latter will be considered in detail when we reach 1882.

But to return to the chronology. The year 1873 produced a weapon which is difficult to classify yet features the characteristics of a surface-running weapon of which, it may be added, there were plenty of examples when inventors realized that they could not devise a satisfactory system to maintain a constant depth if the torpedo ran submerged. Indeed, it was the solution of this problem that formed the basis of Whitehead's success and fortune. The brainchild of the rector of Rye in Sussex, the Reverend Charles Meade Ramus, was described by him in a letter to the *Times*.[2] He explained that the Admiralty had experimented with his "rocket-driven floats" in July of the previous year (1873) and continued: "[They were] calculated to pass along the surface of the water (not through it) at the rate of 80 to 100 miles an hour. These floats will strike a ship precisely at the waterline and, on striking, will sink any ship but an ironclad." And anticipating the Treasury's parsimony, he added, "An apparatus costing not more than £5 will, at the distance of half a mile, destroy any ordinary ship."

The inventor confirmed that he had tested small rockets and floats and had obtained speeds of from forty to fifty miles an hour—which at least suggests that the concept was feasible—and added proudly, "No speed at all approaching this has ever been attained by any floating body before." His claims were apparently supported by the respected naval architect William Froude when he submitted a written report on experiments he had carried out with the floats to the House of Commons.[3]

The reverend gentleman was, however, distinctly sour about the tests which were carried out independently by the Admiralty: "I may add that the torpedo experiments were not made in my presence, and I have every reason to believe were not prosecuted with any great desire to ensure success." It was a complaint similar to those made by Wilson and Ericsson and, some years later, by John Lay. This unanimity suggests that their lamentations had a basis of truth. Unfortunately, no other details of the Ramus rocket torpedo floats are known, as the inventor apparently

did not take out a patent and it would seem that the Admiralty quickly lost interest in this potential one-hundred-miles-per-hour weapon.

Next in chronological order was another of those bizarre curiosities which bestrew the history of torpedo development and help to provide a little light relief. This strange device can be dated to just before 1878 and was designed by Lord Milton, who by the time of the news report in the *Times* was already "the late" Lord Milton. It was described as a torpedo *boat*, but the details given in the newspaper make it impossible to determine whether it was manned or unmanned, although a reference to "two large eyes" in the foresection or head "from which radiate a strong electric light that will exhibit the keel of the enemy's vessel for a considerable distance" suggests that at least one man was accommodated. However, if such was the case, the report remains silent on how he would survive the attack.

On the upper part of the bow was a "powerful ram . . . capable of penetrating an armour-clad." The next paragraph of the report is reminiscent of a passage from Jules Verne in one of his more dramatic moments: "In what may be termed the nostril, there is a revolving gun worked by hydraulic power and fired by electricity . . . [the gun is] rotary, but has four chambers, placed like the spoke of a wheel, so that while one shot is being fired, a second is being charged, a third sponged, and a fourth cleaned so that the shots can be fired in rapid succession." Although the weapon is attributed to Lord Milton, it seems to bear a very close resemblance to Timby's original "revolving battery tower" which Ericsson had simplified for use on the *Monitor*. The explosive utilized by the gun's shells was described as "new," and it was claimed that a single pound of it could displace 137 tons of ironstone. Perhaps Milton, before gaining his place in paradise, had stumbled on the secret of Semtex.

The propulsive power for the torpedo is not clear, though the report states that "it is intended to work under water by compressed air" and there is mention of a propeller at the tail. Similarly the source of the electric power is not revealed while the proposed tactics to be employed during an attack were also somewhat odd. "The boat is sunk to the depth required by taking water in at the bottom, and she could then remain under water . . . from three to nine hours, while in attacking a vessel the speed would be about 18 knots." A scale model some four feet eight inches

in length was brought to the notice of the Admiralty, according to the *Times*'s report, but nothing further was heard of Lord Milton's maritime monstrosity.[4]

An equally unbelievable device was patented by James H. McLean of St. Louis, Missouri, the following year.[5] And as the illustration shows, it was without question the most incredible contraption ever to have been graced with the description of torpedo. Even John Lay in one of his wilder moments could have looked askance at it.

McLean's stated object was to produce a torpedo "which will be cheap in construction," although it is impossible to judge whether or not he succeeded in attaining his aim. He also wanted to produce a weapon which would "automatically attach itself [to the target] and will remain attached . . . until it explodes." A time-delay mechanism built into the detonator provided the necessary interval between impact and ignition of the charge and it is difficult to understand why McLean did not prefer the more orthodox instant detonation with a contact trigger. It is not proposed to waste space on this ludicrous device other than to point out that the warhead (or explosive charge) was attached to the end of a spar that projected from the nose of the weapon, that the running depth was controlled by a series of floats attached by chains to the "torpedo boat," and the preferred method of use required two such boats to operate as a pair linked by a coupling bar. The benefit of such an arrangement is not immediately clear.

When the first experimental VTOL aircraft was demonstrated to an unbelieving group of pressmen it was dubbed "the flying bedstead." Having looked at a drawing of James McLean's torpedo an apt sobriquet for it might be "the sinking bedstead."

The same year, the Frenchman William Giese of Bordeaux patented a rocket torpedo with an anachronistic harpoon-like spear set into its nose.[6] The inventor's explanation of how it worked was, to say the least, a trifle odd. "On arriving in proximity to the vessel intended for destruction the spearhead will not only cut its way through the usual protective netting . . . but will on striking the hull of such vessel, embed itself therein and insure its destruction." As an afterthought the Frenchman added, "If desired, the spearhead may be made the means of igniting the explosive charge . . . in a socket in the nose of the torpedo." Earlier in the specification, he was even more vague about the warhead: "The torpedo is . . .

Most Splendid Results · 89

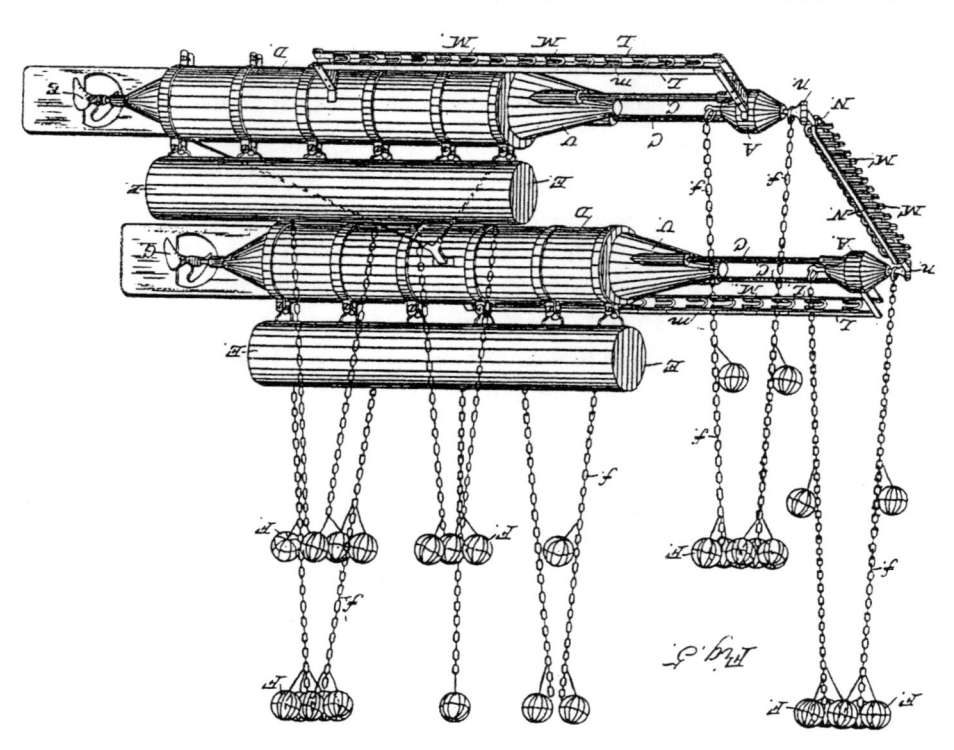

James McLean's strange twin-bodied torpedo is probably one of the most unbelievable underwater weapons ever to be devised, but it was granted a U.S. patent in December 1879.

charged with the usual materials employed for the destruction of vessels at sea or in harbours." It is apparent from the wording of the patent that Giese had neither built nor tested his torpedo at the time he applied for protection. Had he done so, he would have discovered very swiftly that it was unworkable.

Throughout the latter half of the nineteenth century, rocket propulsion was much in vogue with inventors anxious to avoid the complexities and difficulties of compressed-air engines. But depending on one's viewpoint, Capt. John Ericsson's projectile torpedo marked a partial return to sanity by the simple expedient of discarding rocket propulsion and firing it from a gun!

This weapon, although often quoted as dating from around 1880, actually went much further back in time. For in a letter Ericsson wrote to *En-*

gineering in 1870, in which he gave details of his pneumatic torpedo to support his claim that it was superior to the Whitehead, he concluded, "The scope of this mechanical device [the pneumatic torpedo] is but limited. Fully impressed with this fact, my labours were early devoted to plans for carrying on submarine attack by means of which the contest might be removed to the open sea. Before the close of the late [Civil] War, the problem was satisfactorily solved; and during the month of November 1866, the leading features of a new system of naval attack were confidentially laid before the King of Sweden & Norway; the Swedish Minister of Marine, Count B. von Platen, and Commodore A. Adlersparre."[7]

The projectile idea thus appears to have originated before 1866. And a fortnight after this letter appeared, Ericsson placed details of his latest weapon in the public domain by way of a further article.[8] Why the inventor thought so highly of this new device is difficult to understand—but he tended to be overenthusiastic about each new weapon he dreamed of, and, to be honest, he was almost always blind to its defects no matter how obvious they may have been to other people. On this occasion he supported his claims with formulae and mathematical calculations as proof that it worked. Briefly, it consisted of no more than an elongated shell which was fired from a deck-mounted smoothbore cannon—Ericsson giving the gun's caliber as fifteen inches in one paragraph. On being fired, the shell followed a normal trajectory through the air, but when it finally entered the water, it would "not be diverted, but continue to move under the surface, with the same inclination it had on coming into contact with the dense medium."[9] The shell itself was to contain not less than three hundred pounds of dynamite, so it would have provided a loud bang if nothing else.

He gave the range as five hundred yards but appears to suggest that the projectile would be capable of sinking ships at up to five thousand yards. Like so many other torpedo inventors of the period, however, he neglected to say whether he had tested the device successfully before rushing into print. The only recorded trial to have been traced related to a projectile device which was fired from his torpedo vessel *Destroyer*. This will be described later in the chapter.

But to return to the pneumatic weapon, the editor of the *Army and Navy Journal* invited the chief of the Naval Bureau of Ordnance, Commo. William N. Jeffers, to meet Ericsson, and, favorably impressed, he agreed

Most Splendid Results · 91

to an official trial of the "tubular cable torpedo," although the inventor insisted that the government should meet the cost of the air compressor on which it relied for both its power and control.[10] In February and March of the following year, Jeffers reported that the torpedo "worked regularly and without the slightest trouble and to the admiration and surprise of everyone to whom I have shown it."[11] Despite this glowing testimonial, there is no further reference to the "tubular cable" or pneumatic torpedo in Church's comprehensive biography. Presumably the United States Navy had discovered that the weapon was unmanageable due to the drag of its air hose and, in addition, was too slow.

Attention now shifted to Ericsson's projectile torpedo—or as he termed it, "hydrostatic javelin"—and Jeffers provided the inventor with a smooth-bore naval gun which was mounted on a yard boat. The experimental projectile was described as an elongated 15-inch shell some ten feet in length. The tests were to be staged at Sandy Hook, but the two men disagreed about the nature of the gun. Ericsson wanted a revolving turret—shades of the *Monitor* no doubt—while Jeffers favored a simple deck-mounted carriage. To complicate matters there was a change of administration following the election of Rutherford Birchard Hayes, and in the political fall-out further trials were abandoned.

By now, however, Ericsson's agile mind had seized on yet another idea: a torpedo vessel to be called *Destroyer* which would fire its weapon from a submerged tube. Why Ericsson found this arrangement so excitingly novel is difficult to understand, for Whitehead had devised a submerged launching tube as far back as 1868 for the second series of Austrian trials and Thornycroft had built his first torpedo boat in 1873.[12] So neither of his proposals were exactly original.

To judge by his letters, Ericsson's boat was intended to run virtually awash like a Civil War *David*. And during his initial experiments on the Hudson River, he employed a "wooden torpedo" to determine the "form" of the weapon. Encouraged by the results, he proceeded to go ahead with a machine to "project a large torpedo under water."[13] The salient yet strangest feature of this new projectile torpedo was its method of launching. Ericsson first tried compressed air, as favored by Whitehead, though he eschewed rocket propulsion, which other inventors of the period seemed to prefer. He chose, instead, a "propelling piston bolted to [the torpedo's] aft end," which was, in turn, actuated by an explosive charge.

One could detect a hint of sarcasm in Commodore Jeffers's appraisal: "The remarkable simplicity of the arrangement commends it to every practical man . . . which I take to be a reproduction of Sir William Congreve's water rocket."[14]

The biggest drawback to the projectile weapon was its range—a pitiful three hundred feet, only half that of the already inadequate pneumatic torpedo. Even so, at one stage, Adm. David Dixon Porter voiced his opinion that its optimum range should be no more than two hundred feet, at which distance "the projectile could not miss nor the enemy escape."[15] How an experienced and highly competent senior flag officer could imagine any enemy allowing its foe to get as close as seventy-five yards beggars belief. Very sensibly, Jeffers's successor as chief of the Ordnance Bureau took the view that the projectile should have "more range."

It must be remarked that having already usurped Whitehead's credit for building the first practical submerged torpedo tube, Ericsson then failed to acknowledge that his projectile weapon was by no means the first in the field. For, as related in chapter 1, Philip Braham had submitted a very similar idea to the British Admiralty in 1868, right down to the awash-running delivery vessel. But propelled through the water solely by the impetus of its original discharge, and lacking any form of self-propulsion, it suffered from the same obvious weakness of Ericsson's weapon so far as its effective range was concerned. Allowances must be made, however. The Swedish engineer was now close to his eightieth birthday, and age could well have affected his judgment by that time.

The projectile torpedo soon became bogged down in yet another dispute, on this occasion between Ericsson and the *new* chief of the Bureau of Naval Ordnance, who insisted that the *Destroyer* should be tested at sea *and* at the inventor's expense. The arguments bandied between the two men grew more and more acrimonious as it became increasingly obvious that the U.S. government would not finance further experiments. Ericsson's co-investor, the shipbuilder C. H. Delamater, counseled restraint and gently pointed out that the inventor had spent more than one hundred thousand dollars of his own money and devoted twenty years of his life to the project—"thankless public service" was the phrase he used—and urged his friend to abandon the project.[16] Yet, determined as ever, despite his advancing years, Ericsson refused to give up the battle and continued to fight his corner until he died.

The only drawings of the projectile to have been traced do not illustrate the torpedo or the details of the firing mechanisms, although on one occasion Ericsson said that he employed twelve pounds of gunpowder in the form of slow-burning cake powder contained in a cellular cartridge.[17] Bethell also refers to a rubber membrane being stretched across the muzzle of the gun barrel to keep out the water, although this is not apparent in the published drawings which seem to show some form of a hinged cover-plate.[18]

The diameter of the projectile itself was sixteen inches with a gross weight of fifteen hundred pounds while Ericsson personally confirmed the explosive charge as being three hundred pounds. Bethell also mentioned skewed tail vanes[19] to maintain directional stability, but these are not readily apparent in published drawings. However, it is highly probable that Ericsson produced more than one version of the weapon during development. In addition Bethell stated that during tests it covered 310 feet in three seconds–thus giving it a speed of seventy miles per hour. The inventor's own experimental data confirmed both of these performance details.

Perhaps the greatest puzzle rests with the construction of the projectile itself. The *Royal Navy Torpedo Manual* stated that it was "torpedo-shaped . . . [and] . . . made of wood."[20] Ericsson, too, frequently referred to a wooden construction when describing test firings. And the version delivered to the Peruvians–which was, admittedly, to be fired from a surface smoothbore gun–was said to be made from "solid yellow pine."[21] This same report gave a length of nineteen feet eleven inches, whereas both Bethell and Kirby state it to have been twenty-five feet. But in the course of an argument with Commodore Jeffers on the options of bow or beam firing of the projectile which was to be fitted to the *Destroyer*, and which was a development of the earlier model, Ericsson specifically referred to it as having a length of thirty feet.[22] And bearing in mind the source this must be accepted as definitive. But again, there were various versions and it is fairly certain that the surface-fired weapon was smaller in both length and diameter.

Spurned by the U.S. Navy, Ericsson tried to sell the *Destroyer* to the Chinese without success, though he refused to do business with the Russian government when approached by one of the czar's ministers.[23] And his attempt to sell the boat and its torpedo to Peru during the war with

Chile in which Lay's weapon had featured so disastrously was blocked by a U.S. government ban on the export of weapons to the belligerents, although, as noted earlier, one example was delivered by a tortuous overland route but arrived too late to see combat.

The British Admiralty, however, showed a passing interest, and one of the *Destroyer*'s submarine guns and four projectiles were purchased for the Royal Navy. Subsequent trials on a beach in Pembrokeshire on 30 September 1886 did little for Ericsson's reputation, although, like John Lay on a similar occasion, he blamed the adverse conditions under which the weapon was tested as the cause of the failure. Two were fired from a submarine gun suspended below a scow in twenty feet of water, and both struck the bottom from which they ricocheted and then veered to one side.[24] The third, which had been charged with 321 pounds of guncotton, was fitted with a Woolwich-designed detonator in place of Ericsson's percussion lock. The warhead exploded prematurely and, shattering the projectile's powder chamber, damaged the gun barrel beyond repair.[25] As a result, the fourth weapon was not tested and the trial was aborted. Ericsson's biographer blamed the abandonment of further tests to a change of senior personnel at the Admiralty. Such loyalty is commendable, but one has just the faintest suspicion that the disastrous conclusion to the initial trial played more than a small part in the British decision.

The *Destroyer* and its projectile torpedo proved to be Captain Ericsson's swan song. And although still firmly believing in the superiority of his outmoded and impractical weapon, he finally bowed out of the torpedo story.

No precise dates can be attributed to the next three torpedoes in this chronology beyond a vague circa 1880. But all employed steam as a propellant with one being of British origin and two coming from the United States. Unfortunately, virtually no details of the British Peck weapon can be traced. It was described as a steam torpedo, and its inventor, Edward C. Peck, was said to have had a connection with the Yarrow torpedo-boat works, but beyond his name nothing else is known.

The Paulson weapon was, in concept, well ahead of its time. Powered by superheated water, it was kept on course by a bow-mounted compass fitted with electrical contacts that controlled the rudder mechanism. The course was preset before launching. Its main point of interest lies in the

use of its steam motive power to expel water on the ejector principle—an early example of pump or reactor jet propulsion. Little enthusiasm was shown for this novel idea until 1938, when the Torpedo Experimental Establishment at Greenock tried developing a jet-propelled torpedo which, by 1942, could run at twenty-three knots for eight hundred yards.[26] But the propulsive system was considered to be too inefficient and the project was quietly abandoned. The technology is understood to be again under active examination in several countries, but all details are, not surprisingly, classified.

Slightly more is known about the third steam-propelled weapon, the Hall torpedo. This was rather unusual in that its propellant, pressurized water heated to 550 degrees Fahrenheit, was transferred directly into the torpedo from the boiler of the launching vessel. According to Kirby, evaporation of the water under reduced pressure provided a propulsive performance fully equal to the contemporary Whitehead and Brotherhood compressed air engines.[27] This American weapon also boasted an unusual, one might say almost bizarre, roll-control system. Built around a transverse mercury-filled U-tube, the body was fitted with wings that were pushed in and out by the action of the mercury each time the torpedo rolled. These "wings" were angled to provide lift, thus keeping the weapon on an even keel. The Royal Navy Torpedo Manual added that this antirolling device took the form of pectoral fins with air jets on the flank.[28] Hall is reported to have also completely redesigned the Whitehead balance chamber mechanism, though no details of the modifications, or their effectiveness, have been traced. Finally no performance details or other specifications can be established as it seems the weapon never went into production.

Another torpedo to make its appearance during this period was the wire-actuated coastal-defense weapon designed by Louis Philip Brennan and for which a British patent was obtained and backdated to 4 September 1877. This important torpedo and its inventor will be the subject of chapter 8, for he was virtually the only man to successfully challenge Robert Whitehead in the field of self-propelled underwater devices for harbor and river defense.

Passing over the various spar weapons which will be reviewed in chapter 6, the next locomotive torpedo, chronologically, was the Berdan, which, although never precisely dated, seems to have originated in the

early 1880s. For some reason, this device has always been something of a mystery, and so far as naval historians are concerned, the true identity of its inventor has never been previously established. Many acknowledged experts ignore it altogether, and the remainder seemingly share no common ground.

Bethell did not help matters by referring to the torpedo as the Borden, thus starting many on a wild goose chase. He described it vaguely as "an exceptionally large [torpedo] . . . capable of about 2,000 yards at a moderate speed, being steered by tiller ropes from the shore." His further observation, that it was "driven by turbine engines supplied by gas from twelve gunpowder cartridges," suggests that he had some knowledge of the weapon.[29]

Kirby expanded this meager contribution by referring to it as "a rocket propelled floating torpedo which towed a smaller weapon." He also correctly described how the device circumvented torpedo net defenses. The sketch he provided shows the rocket's thrust acting directly on the blades of a turbine, which, in turn, rotated six propellers mounted along the weapon's longitudinal axis.[30] Unfortunately, neither he nor Bethell indicated the source of their information, and it has proved impossible to trace the six-propeller model in any contemporary press reports. Its existence can be neither proved nor disproved.

It is remarkable that although Hiram Berdan was a prominent Civil War soldier and familiar to most military historians, naval writers have, for more than a century, consistently failed to link him to the torpedo, despite brief references to the weapon in specialist works on his army career. But thanks to a recently published biography by Roy Marcot, it is now possible to identify him as the inventor of the weapon that bears his surname, although, even now, some details of the device itself remain uncertain.[31]

Hiram Berdan was born in the small township of Phelps, New York, on 6 September 1824. A mechanical engineer and inventor by profession, he had, by 1860, already been granted six patents, the majority of which related to machinery for the baking industry. In addition, he had another and perhaps more significant claim to fame, having built up a reputation as a crack competitive rifle marksman—nationally acknowledged as "the best shot in the country."

Soon after hostilities began, Berdan used his political connections and

reputation as a marksman par excellence to establish two rifle regiments, the 1st and the 2d United States Sharpshooters. Commissioned as colonel of the 1st Regiment (with Col. Henry Post in command of the other unit), his well-trained riflemen, wearing distinctive green uniforms based on European practice, featured decisively in numerous actions–usually only in company strength, for the Union generals had to make a little go a long way in the early stages of the war. Initially equipped with the Colt Model 1855 revolving rifle, they were subsequently issued with the acclaimed Sharps breech-loading weapon. Indeed, it was boasted that, between them, the Sharpshooters killed more Confederate soldiers than any other two regiments serving with the Army of the Potomac, and they certainly covered themselves with glory on many occasions in both skirmishes and full-scale battles. Always in the thick of the fight, Berdan himself was seriously wounded in 1862 and was left with a legacy of poor health for the rest of his life.

The colonel, something of a maverick, was not popular with either his peers or his superiors, being variously referred to as "unscrupulous" and "totally unfit for command." Major Dyer of the Springfield armory went so far as to brand him "thoroughly unscrupulous and unreliable," and even Marcot, his biographer, described him as "self-serving throughout his life." However, despite this barrage of criticism, Berdan went on to attain the brevet rank of major general–an honor that, despite its temporary nature, he clung to tenaciously for the rest of his career. Then, at the peak of his success, he resigned from the Union army while the conflict was still raging and on 2 January 1864 returned to civilian life.

Free from military discipline, he turned his inventive skills to weapons of war, producing, among other devices, his own repeating breech-loading rifle, an artillery rangefinder, and a special fuse for shrapnel shells which enabled the canister to be detonated at an adjustable predetermined distance from the firing point. He was also involved in the manufacture and supply of rifle cartridges.

Berdan was to spend some seventeen years in Europe. He improved and perfected his repeating rifle in England at the Birmingham Small Arms factory in 1869 and moved on to St. Petersburg later that same year. There seems to be considerable foundation in the claim that his bolt-action weapon formed the basis of Russia's standard army rifle, although Soviet sources denied him any of the credit. Berdan's struggles to gain

acceptance of his various novel ideas by the European military, and a similar failure to interest the United States Ordnance Department in his inventions, fall outside the scope of this book, but he entered the torpedo story in the early 1880s with a remotely controlled weapon which he produced while residing in Constantinople.

A surface device, it ran awash and was steered by tiller lines by an operator who used a sighting mast to follow its course and position after launching. It thus fell within the *dirigible torpedo* classification, for unlike the free-running Whitehead, it remained at all times under human control. An article in the *Scientific American* dated 26 September 1885 gave its speed as twenty-four knots for one hour, although a report in the *Times* of 20 April 1882 had stated it could cover a distance of one mile in ninety seconds—a speed of forty knots. As the weapon was probably not actually built at the time of this earlier report, the performance data cannot be regarded as reliable, and Kirby observed that the "hopelessly inefficient propulsion system failed to drive the weapon forward because the drag of the steering ropes exceeded the [torpedo's] forward thrust."

Intended for use against anchored ships lying behind the protection of steel nets, Berdan's initial weapon towed a smaller torpedo containing the explosive charge astern. When the leading section struck the net, the towed segment "would dive beneath it" and hit the target's unprotected keel—a method of attack in many ways similar to Ericsson's awash boat in which a copper warhead was lowered below the victim's waterline by means of two diagonal rods and an inclined spar as described in the previous chapter. The height of the explosion was one of the recognized weaknesses of the awash-running or surface torpedo, and, as already noted, several inventors had employed analogous mechanisms to enable the charge to detonate below the waterline. Asa Weeks's weapon, described later in this chapter, provided a good example of contemporary thinking.

Berdan's alternative modus operandi envisaged the leading torpedo blowing a hole in the steel net on impact so that the second towed torpedo would pass through the gap and strike the enemy's hull on the waterline. It is unlikely that such optimistic tactics would have been successful in practice, for the blast effect of the first explosion would have probably thrown the second device off course, causing it to miss the hole.

The *Scientific American*'s account of Berdan's weapon yields a few

specification details that help enlarge our knowledge of this particular torpedo, although, as always, the secondary nature of the report means any information must be treated with caution. Its length was given as thirty-one feet, with a maximum width of twenty-one inches and a height of thirty-one inches—proportions which suggest the weapon was not completely cylindrical like the Whitehead and Lay torpedoes. The weight was stated as twenty-eight hundred pounds, with the warhead containing 100 kilograms (220 pounds) of guncotton or dynamite. As already noted, its speed was rated at twenty-four knots.

The article confirms that propulsion was by gas turbines but deviates from other accounts by referring to a thrust of two thousand pounds per square inch from a "compression of rocket powder" rather than rifle cartridges as indicated by Bethell and Kirby. The diagrams throw little light on this aspect, and it is possible that the rocket powder was supplied in cartridge form, which has resulted in a confusion of terms. The sketch certainly fails to reveal Kirby's six tandem-mounted propellers, for it shows only the turbine and a single screw in the stern, forward of the rudder.

As often proves to be the case when unraveling such contradictions, it is possible that Berdan built more than one version of his torpedo and that all three accounts quoted contain a kernel of truth at different stages of development. Before moving on, it should be noted that a report in the *New York Herald* dated 6 June 1886 confirmed that several of the general's torpedoes had been delivered to, and tested by, the U.S. Navy at Willet's Point station, although the highest speed achieved was a meager eleven knots.[32] This contemporary reference at least suggests that the torpedo was actually built and was not, as some have suggested, no more than a drawing-board project. It is clear that further research is needed before a definitive description and assessment of Hiram Berdan's torpedo can be placed on the record.

Having conceived the weapon while living in Constantinople, it was only natural for Berdan to offer it to the Turkish government. And with their traditional inclination to purchase white elephants unwanted by other major powers, they responded with immediate interest. The sultan's ministers, however, were secretly horrified by the general's assessment of their requirements: two bombproof stations each equipped with fifty torpedoes at the entrance to the Dardanelles, plus two hundred

weapons, together with twelve small steamers for the defense of the Ottoman Empire's other harbors. Berdan obviously believed in thinking big.

The financial terms he proposed were even more shattering: £150,000 for the three hundred torpedoes alone–to which would be added the cost of the twelve small steamers and the concrete torpedo stations. He also demanded settlement of an old debt of £12,000 plus interest in respect of a delivery of rifle cartridges some time earlier. Furthermore, "the Turkish government should at once place to his credit the amount required for the torpedoes." The gross amount required was, in fact, actually more than the cash and bullion held by the Turkish treasury, but the *Times* reported enthusiastically, "The Ministry of Finance will have no difficulty in obtaining the money [for] the Galata bankers are quite ready, even anxious, to advance money [against] the revenue of the present year."[33]

Berdan's demands were apparently met, for in a letter dated 18 December 1882, Hobart Pasha, the Royal Navy officer in de facto command of the Ottoman fleet, wrote, "While on the subject of torpedoes I may mention that General Berdan is here proposing a torpedo of his own invention, with which he undertakes to arrive at most splendid results. His Majesty the Sultan has ordered every facility to be given him in the construction of his invention at the Naval Arsenal. When finished, a trial will be made on the Bosphorus."[34] It seems clear from Hobart's phrasing that the torpedo had not yet been built and that Berdan's grandiose sales campaign was based on little more than paper dreams and a modicum of hot air. And it would appear safe to assume that the weapon was not constructed until 1883 at the earliest–a conclusion amply supported by other evidence.

A report in the *Times*, which appeared on 13 June 1882, for example, stated that "the Berdan torpedo project has been for the moment abandoned. The inventor finds, it is said, that the recent improvements in defensive torpedo nets necessitates certain changes in his invention; and that, consequently, the proposed experiments must be indefinitely postponed."[35] From this it must be assumed that Pasha's letter referred to an improved version which had, hopefully, overcome the problem of the earlier weapon. No further reports on the Turkish venture have been traced.

Berdan's later torpedoes, torpedo boats, and even a twenty-four-inch submarine gun all seemed to owe a great deal to other people's ideas.

Most Splendid Results · 101

And confusingly, it is often difficult to distinguish the torpedo from its launching vehicle. An Office of Naval Intelligence information release issued in 1887 refers to a dirigible torpedo but describes a launching vehicle bearing a close resemblance to both Ericsson's awash boat and Asa Weeks's explosive device: "When the boat [*sic*] attacks . . . a long spar projected ahead (of the boat) on striking the enemy reverses the boat's engines and fires the rockets, which then swing (the torpedo) under the net and strike the enemy ship's bottom." In an earlier part of the paper it was explained that "on each side of the (attacking) boat or vessel and some 50 feet from the bow, is a metal tube fixed vertically and opening downwards, large enough to hold a single rocket torpedo capable of carrying 200 pounds of dynamite. Each torpedo is connected by a wire rope to a stout bumpkin forward, and on being discharged downwards will swing forward around it as a center."[36] The syntax may be convoluted, but the broad principles seem clear, and the description given in this official paper is almost identical to Miles Callender's "marine torpedo" patent of 1862, the details of which will be found in chapter 1.[37] Indeed, the similarity is so striking one wonders whether Berdan had "borrowed" Callender's idea. It must also be assumed that this plagiarized design superseded the earlier version with the towed explosive charge.

In this connection, it is interesting to note that a report in the *Scientific American* of 21 May 1887 refers to "improved torpedoes" with an overall length of eight feet and a fourteen-inch diameter which were armed with the same two-hundred-pound dynamite charge mentioned in the Office of Naval Intelligence's information release. Berdan proposed to fit this apparatus into a 150-foot boat, but despite the backing of Adm. David Dixon Porter, the concept was rejected by the U.S. Navy's Torpedo Board.

Undeterred, Hiram Berdan set about building an even bigger vessel, which he christened the *Berdan Ironclad Destroyer* and submitted it to the Navy in January 1892. Said to displace twenty-four hundred tons with twin screws and a maximum speed of twenty knots, it was remarkable for the 24-inch submarine gun mounted below the waterline in the bows—a weapon for which Berdan obtained U.S. Patent 478,215. The similarity to Ericsson's *Destroyer,* with its submerged, bow-mounted launching tube for his unpowered projectile torpedo, described in the previous chapter, will not pass unnoticed by the reader. This, too, it will be recalled, owed a great deal to Philip Braham's original concept dating back

to 1868. If Berdan showed a similar inclination to steal other men's ideas during his heyday in the Union army, it is not surprising he was disliked by all and sundry.

Somewhat curiously, Admiral Porter had also spoken up in favor of Ericsson's projectile torpedo—advocating a reduction in its already inadequate combat range. Support by the admiral seems to have been the kiss of death for all the torpedoes with which he became involved. For like Ericsson, the entrepreneurial former Union general had his ambitious proposals rejected yet again by the United States government.

The ironclad destroyer proved to be Hiram Berdan's final venture into the arcane world of underwater weaponry. He died suddenly from a heart attack while visiting his Washington club on 31 March 1893 and was buried, as befitted his rank, with full military honors in Arlington Cemetery.

As the technology of the Whitehead and Woolwich torpedoes grew apace, inventors produced some astounding freaks in their efforts to keep up. And a surprising number turned back the clock and ignored the basic automobile torpedo design as first envisioned by Robert Whitehead. A fully submerged method of attack was clearly preferable to a surface or awash-running weapon, but many chose to overlook the obvious advantages of the former. This was partly due to the continued insistence on the part of some experts for a controlled torpedo that could be steered toward its target and therefore needed to be visible to the operator. But it was also due to the unadmitted failure of inventors to produce a workable mechanism for depth-keeping, even though by that time Whitehead's system, the Secret, was common knowledge among the cognoscenti.

Another inexplicable blind spot was the failure to develop a self-contained propulsive system independent of hoses, wires, or cables—for once the latter were involved it was almost impossible for them to be fired from tubes or otherwise used by ships at sea. This problem still bedeviled weapons right up the close of the century—a notable example being the Sims torpedo, the trials and tribulations of which will be examined in a later chapter. Whitehead had shown the way, but many inventors chose to ignore the obvious advantages of self-containment and continued to follow their own, usually unsuccessful, paths. Nevertheless, much more work was needed to perfect the fish torpedo. And the advances made in the following two decades—the adoption of the gyroscope for the stabilization of depth and steering control, and the enhanced performance re-

sulting from the application of the heater principle to its compressed-air power units, turned it into a superlative and lethal weapon which, when employed by that other contemporary invention, the submarine, was to bring Germany to the verge of victory in the Great War's ruthless yet very effective U-boat campaign.

A prime example of mixing new-style technology with, by then, outmoded concepts was Asa Weeks's rocket-powered spar torpedo. Although basically a spar weapon, Weeks's device enters the chronology at this point because it is usually classified as a rocket torpedo. And once again the ever-reliable Peter Bethell provides some vital details regarding its performance, though he gives no indication of his source.[38]

He describes it as a surface-running weapon driven by two rockets with a claimed range of one hundred yards at a speed of over forty knots. He also adds with characteristic dryness, "Its direction was a matter of luck." Weeks took out several American patents for the weapon but only one has so far been traced,[39] and this related only to the bow section and the spar mechanism. Weeks mentioned a pending patent application for an "improved" torpedo and this, when found, might clarify the overall design. To confuse matters further, he describes the explosive charge as a "torpedo" but labels it as a "rocket" in his drawings.

By combining the patent with Bethell's brief description the Weeks torpedo emerges as a surface-running weapon, propelled by two rockets, with its warhead mounted on a spar in the bows, and somewhat similar to the rocket float devised by the Reverend Charles Ramus in 1873. On striking the target, the spar was forced back in an action that released the explosive charge. The line linking the charge with the spar was thus pulled taut as the warhead dropped down and this movement activated the detonator. In that short interval of time the charge had sunk beneath the surface and the concussive force of the subsequent explosion struck the enemy vessel below the waterline. As an alternative option, the charge could also be fired by electricity.

Why a straightforward pivoting spar, stripped of such complications, was not to be preferred remained unexplained by the inventor. And John Lay would have surely approved the torpedo's "extreme simplicity." That Asa Weeks probably never built the contraption is evidenced by two items in the patent. First, no model was provided for examination. Second, the Minneapolis inventor stated that his patent was "a full clear and

exact description of the invention, such as will *enable others skilled in the art to which it appertains to make and use the same*" (emphasis added). It therefore seems very unlikely that the Weeks torpedo ever got beyond the drawing board—a fate shared by so many other nineteenth-century devices.

Another rocket torpedo made its appearance in the same year—this time from the hallowed workshops of the Royal Laboratory at Woolwich. It is mentioned in the *Royal Navy Torpedo Manual,* and unlike the Weeks, this weapon was actually built and tested, although, unfortunately, it "only produced a run of 50 yards at low speed."[40] It was said that the design was the work of a foreman at Woolwich and that he had defrayed the costs out of his own pocket. No name was given, but as only ten years had passed since a Royal Laboratory foreman, Lowe, had successfully modified the first Whitehead torpedo produced at Woolwich, it is tempting to wonder whether he was still working at the arsenal and had been the progenitor of this particular weapon. The *Torpedo Manual* concluded dismissively, "The idea was not proceeded with."

In 1884, an English journal provided details of another electrically propelled torpedo designed by an American, J. S. Williams, who was resident in London at the time.[41] This appears to be the first torpedo to employ electricity for propulsion; the earlier Milton weapon had run on compressed air and only used electric power for subsidiary functions such as lights and so on. There are also good reasons for regarding it as the world's first truly electric torpedo, for the article refers to the use of accumulators for storing energy in what were termed "floating magazines," which, in turn, drew their power from a high-tension grid system ashore and to which they were linked by cable.

The torpedo, too, was equipped with an umbilical electric cable, but this seems to have been solely for the purpose of controlling the steering by a keyboard operator either ashore or in one of the floating magazines. Williams even envisaged controlling the weapon from an attendant ship or launch.

As confirmation that it was a true electric torpedo the journal's report added, "The accumulator or secondary battery is specially designed to store energy in a small space and with a light weight, and so as to command an amount of energy representing the power necessary for a speed of 25 miles an hour or more." The range was limited to the length of the

Most Splendid Results · 105

controlling cable which was said to be approximately one and a half miles.

The device was designed for the purposes of coastal and harbor defense and was claimed to be superior to John Lay's torpedo because the "use of such a medium as gas or air under compression . . . is attended with other serious difficulties and disadvantages." Unfortunately, and so far as can be ascertained, the Williams electric torpedo was never built or tested—a great pity, for the weapon was seemingly based on sound principles and operational practicality and was certainly a forerunner of today's wire-guided electric torpedoes. Its unsuitability for tube launching appears to have been its only disadvantage—and not an important one for a weapon primarily intended for coast and harbor defense work.

CHAPTER 6

Filled with Cork and Glue

"I DON'T KNOW what they do the enemy," the duke of Wellington once remarked of the English foot soldier. "But, by God, they frighten *me.*"

It was a sentiment no doubt shared by many young naval officers as they went into battle with a primitive spar torpedo projecting menacingly over the bows of their cockleshell commands. And it was a justified fear, for it was a weapon that probably killed more of the gallant men who served it than it ever did of the enemy.

The spar torpedo, as the name suggests, was merely a canister of explosives fastened to the end of a long wooden pole which was poked against the side of an opposing ship and, when contact was made, detonated. It gained in sophistication with the passage of time and ultimately boasted electrical firing circuits, but it never entirely eliminated the danger of self-immolation. It was, in essence, the naval equivalent of a medieval cavalry lance—but considerably more terrifying to friend and foe alike with its predilection to kill and maim indiscriminately.

The idea probably originated with Robert Fulton and there is evidence of the weapon's use during the war with America in 1812. One spar torpedo boat was seized and sunk by HMS *Ramillies* off Long Island in 1813, and other attacks, especially those on the Great Lakes, are shown in contemporary records. However, no Royal Navy ships were sunk and little damage was caused. Underwater weapons such as the spar torpedo and mines were regarded as inhumane and ungentlemanly by most naval officers and were given no official encouragement until the outbreak of the American Civil War, when the Confederacy, overwhelmingly weaker than the Federals when it came to defending its coasts and rivers, turned to the development of more unorthodox methods of protecting itself from the depredations of the Yankee navy.

[106]

Filled with Cork and Glue · 107

Care must be taken to identify precisely the weapons used at this time, for, as pointed out previously, until the advent of the Whitehead, all underwater weapons were given the generic title of "torpedoes," a catch-all word that included, in particular, various types of both static and floating mines. And so, when the Confederate Torpedo Bureau was set up at Richmond under the command of Brig. Gen. G. J. Rains in October 1862, its main interest was in static devices such as mines and other forms of underwater explosive charges. Gradually however the thoughts of the Bureau's officers turned to the offensive use of such weapons and numerous variations were developed: drifting torpedoes, explosive canisters that were floated downstream, usually at night, suspended from floating logs; or driftwood raft torpedoes that lurked on the surface to trap the unwary; and of course the traditional "infernal machines" that were swept down on enemy ships by the tide or a riverine current.

The spar torpedo evolved from this motley collection of devices produced by the bureau, and Gen. Pierre Gustave Toutant Beauregard attributed its invention to an officer of the Engineering Corps, Capt. Francis D. Lee. Many of these spar weapons were abandoned by the Confederate forces when they were forced to evacuate Charleston and Richmond. The explosive canisters varied in shape and size, as did the material from which they were constructed. But all were intended to be used at the end of a wooden pole or spar projecting over the bow of a "torpedo boat." According to one source, this spar was attached to the vessel by a goose neck fitted to a socket bolted to the bow, near the waterline.[1] Guys from the spar to one side of the vessel kept the spar in its position when the torpedo was submerged for an attack, and it was raised or lowered by tricing lines and tackles.

Multiple chemical fuses—the numbers ranged from five to seven—and contact triggers served to detonate the powder-filled charge, which could weigh up to fifty pounds, while the spar itself was generally around twenty-five feet in length. This meant that the crewmen of the attacking vessel, exposed on deck and unprotected by armor, were barely thirty feet from the explosion, a proximity which many regarded as virtually suicidal.

Lee's device so impressed Lt. William T. Glassell that he tried to persuade his superiors to equip a flotilla of boats with spar torpedoes and dispatch them to attack the blockading Union fleet lying off Charleston.[2]

Failing to obtain official approval—and more important, public funds to meet the costs—he fitted out some small rowing boats with the aid of private finance supplied by George A. Trenholm and gathered together an intrepid group of volunteers. But despite now having the tools, Glassell was again denied permission to launch an attack with these "new-fangled notions" by Commodore Ingraham, the flag officer at Charleston, this time on the spurious grounds that his rank did not entitle him to command more than one boat. So, taking the commodore at his word, Glassell proceeded to carry out the attack with a single boat and a crew of six men. He certainly deserved to win honor and glory, but on this occasion he was to gain neither. For approaching the Union frigate *Powhatan* an hour after midnight, and while Glassell was lining up the boat for an attack, one of the men mishandled his oar either deliberately or in a sudden moment of panic. Losing their rhythm, the other rowers were thrown into confusion and Glassell was forced to abandon the assault, cut free the torpedo, and hastily return to Charleston to face the unconcealed scorn of his senior officer.

As the war progressed, rowing boats were replaced by double-ended steam vessels known as *David*s, most of which were constructed of iron and designed to run awash. Between forty and sixty feet in length with a single smokestack, each boasted a small steam engine in the after part of the boat with a boiler situated in the forward section. Between these two came a tiny cubby hole in which the captain and engineer sheltered with the crew. The twenty-five-foot-long spar could be swung down so that the sixty-pound charge of grain gunpowder would explode below the target's waterline, the detonator being either activated by a percussion trigger or a chemical fuse in which a glass vial of sulfuric acid was broken over an amalgam of potassium chlorate and white sugar.[3]

Despite his lack of success against the *Powhatan*, Glassell was rewarded with command of a new *David*-type torpedo boat—the first, in fact, to be completed—and on 5 October 1863, he carried out an attack on the Union frigate *New Ironsides*, a ship that had already survived undamaged from an earlier attempt to sink her by a small launch which had run into the warship's mooring chains and was forced to retreat with unseemly haste. Glassell, in fact, nearly suffered a similar fate when the *David* was spotted by the frigate's officer of the deck, but, snatching up a

Filled with Cork and Glue · 109

musket, he mortally wounded the Union lieutenant and slammed his spar into the enemy vessel before any further alarm could be raised.

The torpedo exploded and, as the frigate reeled under the blast, wreckage and water cascaded down on the *David* like a biblical plague, extinguishing her boiler fire and jamming the machinery of the steam engine. Fearing the worst Glassell ordered his men to abandon ship before diving into the blackness of Charleston Bay himself. Union sailors dragged him from the water an hour or so later and took him prisoner.[4] The fireman, James Sullivan, was also captured but the other survivors swam back to the torpedo boat which had miraculously remained afloat despite its badly damaged condition. The engineer, James H. Tombs, succeeded in relighting the boiler and, after carrying out emergency repairs, they were able to crawl back to Charleston a few hours later, their boat riddled with bullet holes as testimony to their valor. *New Ironsides* was badly damaged by Glassell's torpedo and, forced to withdraw from the Union blockade squadron, took no further part in combat operations until January 1865.

It is interesting to note that the first hand-cranked submarines were developed from the *David* design and these too, somewhat incredibly, were armed with a spar torpedo in their bows. The most famous, *Hunley,* which sank after destroying the corvette *Housatonic* on the night of 17 February 1864,[5] was raised from the bottom of Charleston Bay in the vicinity of Sullivans Island in August 2000 and is now being examined by marine archaeologists at a conservation laboratory situated in the old Charleston navy base.[6]

The most famous spar torpedo attack carried out during the Civil War, made by Lt. William Barker Cushing against the Confederate ram *Albermarle,* took place on 7 October 1864, using the weapon invented by John Lay which was described in chapter 3. The target was lying off Plymouth, North Carolina, and Cushing had to negotiate eight miles of the enemy-controlled Roanoke River in pitch darkness to reach his quarry. It was a hazardous undertaking even without the additional danger of working with a new and untried weapon. The initial plan, however, did not envisage sinking the ironclad but capturing it in what was known in the nineteenth century as a "cutting-out expedition." And for that purpose two steam launches, each forty-seven feet long with a beam of nine and a half feet, were brought down from New York. One was lost on the voyage

south but the other arrived safely and, armed with a 12-pounder howitzer and a Lay spar torpedo, was placed under the command of twenty-one-year old Lieutenant Cushing—an officer who, despite his youth, had already earned an enviable reputation as a brave and determined fighter.[7]

The first attempt was abandoned when the steam launch ran aground, but the following night, 27 October 1864, Cushing proceeded upriver in heavy rain with his fifteen-man volunteer crew, with a small cutter in tow astern. The plan was to seize the Confederate picket guards with a party of men from the cutter and, having thus prevented the alarm from being raised, run the torpedo launch alongside *Albermarle* so that another party could board her, overcome the ram's crew by surprise, and take their prize downriver to Union-held territory. The enemy ship was known to have only a small crew on board, and Cushing was confident that he could achieve his objective despite the odds. In any case, it was a gamble worth taking.

Luck often favors the brave, and the first part of the plan presented no problems. In fact, the pickets on the river bank failed to observe the stealthy approach of the torpedo launch, and rather than risk the chance of discovery, Cushing decided to discard phase one, seizing the party of Confederate troops on the river bank, and to proceed directly to phase two: boarding and capturing the *Albermarle*. Rounding the final bends of the Roanoke in the southern approaches to Plymouth, he found that the fires used by the enemy to illuminate the water and guard against a surprise attack, were now little more than smoldering heaps of ash and, with increasing confidence, he changed his mind again and steered for the bank to disembark the landing party. But suddenly everything went wrong. While the sleep-heavy eyes of the guards may have failed to detect Cushing's two-boat flotilla in the darkness their dog proved to be more alert. And the animal's furious barking quickly aroused the slumbering soldiers. Fuel was hastily added to the fires and, as shots were fired and challenges shouted into the darkness, the soaring flames of the watch-fires leapt into the night sky to reveal Cushing's party moving slowly upriver.

If the expedition was to succeed, the Federals had only one weapon left: John Lay's untried spar torpedo thrusting ahead into the gloom that lay beyond the bows of the leading boat. Cushing did not hesitate. Calling for full power from the engineer manning the diminutive steam engine,

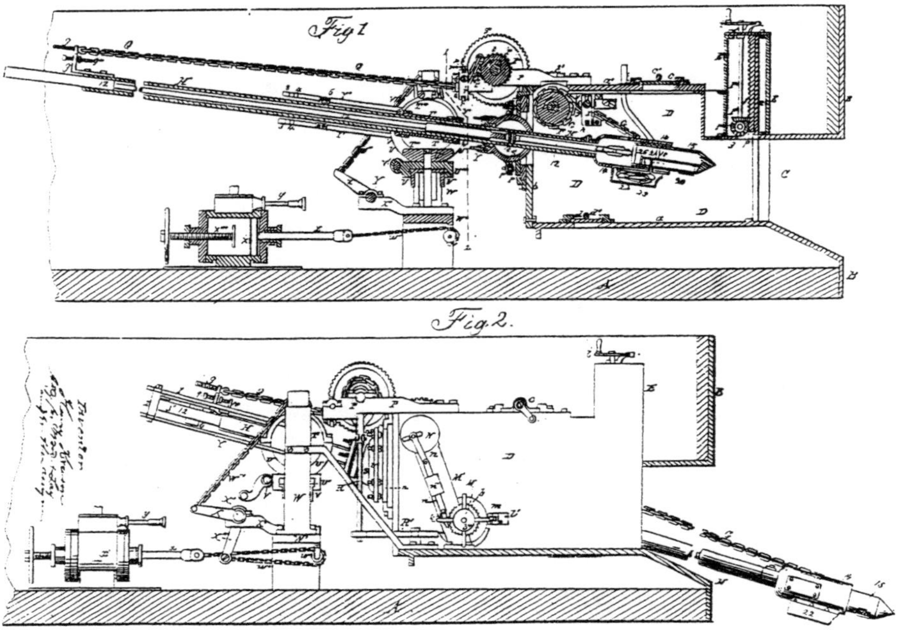

Filled with Cork and Glue · 111

The amazingly complex mechanism of the Lay-Wood spar torpedo of 1865. This was an "improved" version of the weapon employed by Cushing to sink *Albermarle*.

he ordered the helmsman to steer for the *Albermarle,* now silhouetted against the skyline by the hungry flames of the bonfires. Suddenly, in the final seconds of the attack, Cushing realized that the ironclad was protected by a defensive log boom, and on full reverse helm the launch altered course just in time to avoid disaster. He cautiously retreated a few hundred yards downstream and then turned the bow of the little steam launch toward the enemy again hoping that, this time, the longer run at maximum speed would enable him to jump the logs and thrust the spar into *Albermarle's* keel.

He was met by a hail of bullets from both the troops on the river bank and the men lining the rails of the armored ram, but with Old World chivalry, he shouted to the Confederate sailors, "Leave the ship! We're going to blow you up!"–a warning reinforced by the flash of powder as his solitary deck-mounted howitzer lobbed a shell at *Albermarle.* Seconds later, the launch jumped the log barrier with the grace of a salmon

returning to its spawning grounds. Using the guide ropes, and coolly remembering in what order to employ them, Cushing lowered the spar and, as the explosive canister struck the enemy's keel, jerked the lanyard to detonate it.

The roar of the explosion echoed across the river, a tall column of water rose high into the air before crashing back and drenching both attacker and victim, and "*Albermarle* heeled visibly whilst the water rushed into her."[8] But like Wellington at Waterloo, it had been "a close run thing," for at the very moment the spar touched the enemy's hull a gunport opened directly in front of the Union launch to reveal the menacing black muzzle of a loaded 100-pounder. It was barely thirty feet from its minuscule opponent, and a bloody death faced every man aboard Cushing's gallant command as an unseen hand closed the firing lock. The blast of the exploding torpedo lifted the Confederate vessel out of the water and in that brief miraculous moment the gun fired and the shell shrieked safely over the top of the launch. Not that Cushing's boat escaped scot-free from the melee. The fountain of water that erupted when the torpedo detonated had put out the boiler fires while the wreckage and debris falling from the sky disabled its steam engine. It was an almost identical situation that had faced Glassell when he attacked the *New Ironsides* and, like his Confederate counterpart, Cushing took the same decision and gave the order to abandon ship.

Utter confusion now reigned, but amid the smoke and flames the crew of the torpedoed ram could be heard calling on the Union sailors to surrender. A few did so and were taken prisoner; others remained in the water until overcome with exhaustion before being pulled out of the river and taken ashore to share the same fate as their comrades. Cushing, meanwhile, had succeeded in making his escape, and, having swum several hundred yards downstream, he met up with the launch's only other survivor. He had been badly wounded, however, and, with his own strength failing at every stroke, Cushing was forced to leave him to the mercy of the black waters of the Roanoke. On finally reaching the shore the young Union lieutenant hid in a swamp throughout the next day and, the following night, having found a boat, returned safely to rejoin his squadron. Two men had died. The rest had been taken prisoner. He was the only survivor of the epic attack.

On this occasion the spar torpedo had done everything that could be

expected of it and in this David and Goliath contest, the giant had been slain. But its success was probably more due to Cushing's daring bravery than John Lay's engineering skills, for, as usual, his design was far too complicated. Military writer H. W. Wilson observed that Cushing "had four various strings fastened to his hands or feet, and if any one of these had been pulled at the wrong time, the attack must have miscarried."[9] Bethell picked up the same point and his barbed comment says it all: "If anything more be necessary for the attribution of this weapon to Lay alone, it is that Lieutenant Cushing had to have four separate strings tied to his hands and feet, by which to operate it."[10]

It is hardly surprising that William Barker Cushing has become one of the United States Navy's heroes. Sadly, he did not live to enjoy his glory into ripe old age, for he died in a Washington government hospital on 17 December 1874, only ten years after his triumphant attack on the *Albermarle*. He was just thirty-two years old.[11]

Although General Beauregard had credited Capt. Francis D. Lee with inventing the spar torpedo, a patent had been granted to Isaac A. Ketcham of Brooklyn, New York, on 14 October 1862 for "an apparatus by means of which an explosive shell . . . may be projected from the side, bow, or stern of a vessel in any direction of position under the surface of the water, and exploded while it is held, or after it has been left beneath the enemy's vessel . . . which it is desired to destroy."[12] The drawing appended to the patent specification depicts a typical pivoted spar torpedo being operated on the vessel's beam. The application date for the patent is not shown but it must have certainly been before the Confederate Torpedo Bureau was established in October of that same year. However, while Lee actually built his torpedo, Ketcham's patent makes it clear that he was only putting forward a theoretical concept—as witnessed by the conventional phrase "To enable others skilled in the art to which my invention appertains to fully understand and use the same." It is a sentence which will be encountered many more times in the course of this survey, and it is as well to appreciate what it means.

The distinction is really only academic because neither man was the true inventor of the spar torpedo. At the beginning of this chapter it was observed that the concept "probably originated with Robert Fulton" and there is strong evidence to show that he was the first to come up with the idea. First, however, it is necessary to dispose of the parallel claim that

Fulton and his associate Joel Barlow had also built and tested a crude self-propelled torpedo before 1797. And that, during experiments on the River Seine, something had gone seriously wrong and both men had nearly lost their lives as a result. But no reliable evidence has been adduced to support the story which must be regarded as apocryphal.

There are firm grounds for accepting Fulton's responsibility for the spar weapon. And it is on record that on 4 September 1813 he "described a radical new method of "torpedo" attack to [Capt. Stephen] Decatur. Fulton drew a diagram of a friendly vessel supporting two very long spar torpedoes several feet below its keel, which "were to be rammed under the hull of the enemy vessel."[13]

It seems probable that Fulton never proceeded seriously with his spar torpedo but as he was involved in an advisory capacity during the war with England in 1812–13, and in view of his continuous interest in all forms of underwater weaponry, the attack on HMS *Ramillies* off Long Island and similar assaults in the Great Lakes referred to at the beginning of this chapter, may well have originated in his fertile mind. The "other" torpedoes attributed to him by historians were mainly in the nature of drifting mines, although his submarine gun, "Columbiad," of 1813 bore an uncanny resemblance to Ericsson's later smoothbore weapon of the same generic name. And his use of the word "bullets" in connection with the gun suggests that he had in mind an elongated projectile similar in essence to that employed by the Swedish inventor.

Fulton's towed torpedo of 1807 will be examined in the next chapter along with other weapons of that type.

The year 1862 witnessed two new variations on the spar torpedo theme. The first was a fairly simple form of a swiveling boom invented by William H. Elliot of Plattsburg, New York.[14] Terming it "a submarine battery," Elliot described it as being "the employment of a magazine attached to an arm extending from a vessel underwater, [the] said arm being attached to the vessel by means of joints, so that it may receive either lateral or vertical motion"–an idea, in essence, very similar to that patented by Isaac Ketcham later the same year which was mentioned earlier in the chapter. The explosive charge was triggered by electricity. The specification is unusually brief and not much more can be added to the bare bones outlined above, but the drawing that formed part of the patent suggests that "the vessel" ran awash, and there is reference to breast-

Filled with Cork and Glue · 115

works "to protect the men from rifle ball while charging the arm." It is also apparent from the text that this was a paper project and was never actually built or tested.

Oliver C. Smith from Salem, Massachusetts, produced an altogether more complicated device which he dubbed an armor-clad.[15] Armored, as indicated by its name, it was steam-propelled with a bombproof upper deck. Armed with a revolving gun turret it also had two "horns" at each end of the hull which were, in effect, spar torpedoes. To these Smith proposed to add mortars "or other pieces of ordnance or explosive contrivances . . . [to] . . . work between them." It was, again, a purely paper project, and the drawings do not inspire any confidence that it would ever have become a practical weapon.

It is difficult to define precisely into which category the next device falls, a locomotive or a spar torpedo, as it would appear to be a combination of both forms. It was patented by James D. Willoughby of Washington, D.C., in 1864, and while its upper section, which ran on the surface, was rocket-propelled, there was an alternative system of rocket propulsion in the submerged section suspended beneath the main body, for use in night attacks when the surface flames would reveal the weapon's approach.[16] This submerged rocket unit was part of a hollow sparlike tube which carried an explosive charge at its forward end and this was arranged so that the "torpedo" (or charge) would have struck and destroyed its victim before the surface section was visible. Although ingenious in concept—Willoughby called it a "travelling torpedo"—it is highly doubtful that the submerged rocket would have worked. And whether it ever, in reality, lived up to its inventor's optimistic description must remain a matter of conjecture.

The following year (1865) produced a genuine spar weapon designed by George M. Ramsay of New York City, though the patent also covers the torpedo boat on which it was to he mounted.[17] Cutting through the verbiage of the specification this was really no more than an elaborate, and seemingly unnecessary, refinement to the method of detonating an explosive charge. On striking the target the warhead drove an iron rod back through the hollow shaft of the spar and, to paraphrase the inventor's words, it tripped the cock of the lock, burst the cap, with a resultant instant explosion. While it is easy to ask why a simple contact or chemical detonator would not have been more effective, it is noticeable that many

116 · *Nineteenth-Century Torpedoes and Their Inventors*

of the single-detail patents taken out during this period—and which, for reasons of space, have been excluded from this narrative—relate to improvements in detonator and trigger mechanisms. Only one reasonable conclusion can be drawn from this persistent interest by inventors: spar torpedoes could not be relied upon to explode every time they struck their target. Despite advances in technology during the intervening period, a very similar problem was to bedevil both the American and German submarine services in World War Two.[18]

Scovil S. Merriam from Springfield, Massachusetts, came up with a fearsome early submarine armed with a spar torpedo in 1866.[19] It is not proposed to examine the submarine section of the patent as it falls outside the parameters of this volume; however, its lack of hydroplanes and its rather doubtful reliance on water and heavy weights for ballast control does not encourage optimism in the vessel's seaworthiness. And it is perhaps fortunate for Scovil Merriam that he never actually built or tested his submarine, for while it would have no doubt submerged successfully, its return to the surface would have been rather more problematical.

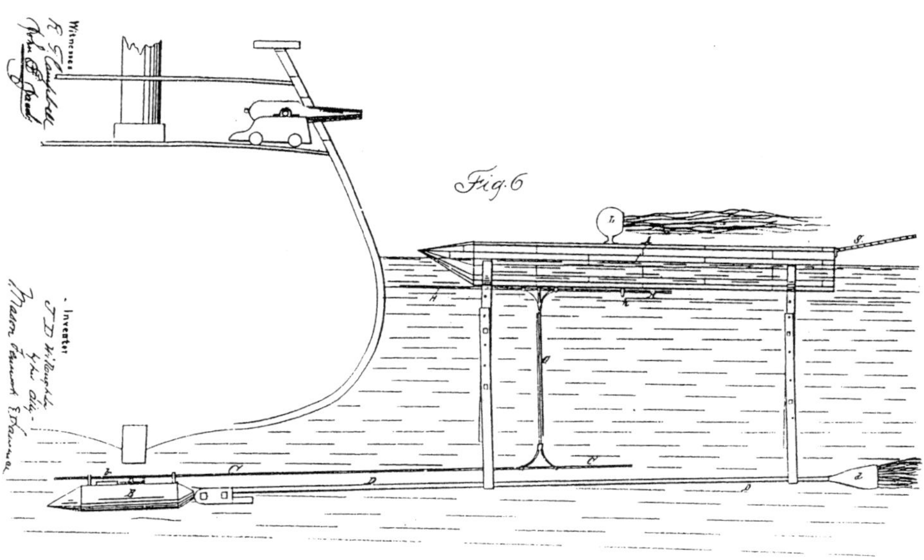

Willoughby's rocket-propelled float torpedo, 1864.

His spar torpedo was simple but probably equally suicidal in operation. It comprised a bar, or spar, presumably made from iron to judge by the text, to which an explosive charge was attached. This bar was extended forward through the bows mechanically but Merriam gave no indication of its length nor what he considered a safe firing distance. The drawings with the patent were not scaled but it seems likely that the spar was around twenty feet in length which, if correct, would have made the explosion dangerously close to the submarine. Not surprisingly no one appears to have taken up the idea.

The executors of the late Charles J. von Court of Morrisania in New York state, were granted a patent for what appears to have been a retractable spar.[20] For added interest the inventor also envisaged a twin spar version. The idea was to keep the spar inside the launching vessel until action was imminent. To achieve this, the rod on which the charge was fastened was fitted inside an inboard tube which was filled with packing to keep it watertight. The rod was thrust forward by means of a hand-cranked rack and pinion mechanism which could, of course, also be reversed to draw the spar back into the vessel. There was provision for electrical firing and, if this was to be employed, the insulated wire carrying the current was threaded neatly through the hollow tube to the charge, thus protecting it from damage by enemy fire and the danger of water penetrating the insulation to cause a short-circuit.

Von Court's drawing showed only a single tube and spar, but his specification added, "It is evident that a plurality may be used and that, instead of the rod passing from the bow of the vessel, it may pass from the side or sides thereof, or both, if required and found necessary." Except for Fulton's design study of 1813, this is the only known example of a twin-sparred weapon. Von Court's proposed tactics were, however, a little curious, for he suggested that the spar and its charge should be forced into the enemy's hull and "remain there while the torpedo-carrying boat is propelled to a safe distance [before] the torpedo [is] fired." Yet examining the drawings and text of the patent, it must be concluded that the apparatus might have worked if an enterprising entrepreneur had taken the time and trouble to develop it.

Very few details have survived the Uhlan ram torpedo which formed the front end of a seventy-foot-long torpedo boat built in Germany around 1876. The "ram" was said to be ten feet in length and the "tele-

scopic forepart [of the vessel was] filled with cork and glue."[21] From the description given it was probably more in the nature of a detachable explosive ram device than a conventional spar torpedo.

There is, unfortunately, a repetitive and deadly dullness about most spar torpedo inventions. Although, to be fair, there are really only a limited number of ways to modify or improve the weapon's basic form. Yet its combat use against the enemy was frequently one of the most exciting means of destroying the foe dreamed up by inventors in the nineteenth century. Examples of the spar torpedo in action during the Civil War have already been described, and only a year after the Uhlan's appearance, the weapon found itself engaging the enemy once again during the 1877–78 Russo-Turkish War. This time, to make matters more interesting, the spar faced direct competition from both the self-propelled Whitehead and, on three separate occasions, the towed torpedo. Which would emerge as the superior weapon in the heat of battle?

Urged forward by public opinion, the Russian government declared war on its traditional enemy, Turkey, on 24 April 1877, with the ostensible purpose of supporting the oppressed Christian population in the Balkan peninsula, an area at that time part of the Ottoman Empire. Most of the naval operations centered on the river Danube, where a Turkish flotilla maintained control of the great eastern European waterway which blocked the route of a Russian land advance to the south. The maritime war also extended into the Black Sea, which both countries sought to control.

The czar's troops began crossing the Danube on 23 June at Zimniza, and by the following day, work had already begun on the construction of a thirteen-hundred-yard army bridge at Nicopol. On land it was a see-saw campaign in which first one side, and then the other, gained ascendancy. Kars finally fell to the Russians on 18 November 1877, and an armistice was granted at Andrianople on 31 January the following year. The Treaty of San Stephano, signed on 3 March, released Bulgaria, Serbia, and Montenegro from the Turkish yoke. But the settlement threatened to upset the balance of power in the Balkans and, headed by Britain, the Great Powers of Europe met for the Congress of Berlin in the summer of 1878 to redraft the treaty. Under this revised protocol, Bulgaria became an autonomous principality under Turkish suzerainty and a new state, Eastern Rumelia, was created under the political and military rule of the sultan. Bosnia and

Filled with Cork and Glue · 119

Herzegovenia were joined to the Austro-Hungarian Empire, while Montenegro alone gained true independence. It was certainly not a case of to the victor the spoils! And the political and ethnic fallout from this concerted effort to prevent Russia gaining too strong a foothold in the peninsula was to result in the dangerous political instability that continues among the Balkan nations even into the twenty-first century.

Such was the background to the war. Let us now examine the part played by the torpedo in the conflict.

The first clash came on the night of 12 May, when four Russian launches armed with towed torpedoes attacked an unidentified Turkish warship lying at anchor off Batum. Only one torpedo was apparently employed and it proved to be tactically accurate. But although it successfully snagged its prey, the electric firing key failed to trigger the detonator and the attack was aborted. In the words of the old proverb, you can take a horse to water but you cannot make it drink.

There was a similar disappointment in June when another towed torpedo attack at Sulina, this time against the corvette *Idjilalie,* failed for an identical reason. Finally, on the night of 23 August, an assault mounted by four Russian steam launches against the corvette *Assar-i-Chevket* while it was anchored off Sukhum Kale led to a claim of two hits. But, in fact, two torpedoes had exploded well clear of the target while the third, despite striking the corvette, failed to detonate and inflicted little damage to the warship. One launch, *Torpedoist,* failed to return.[22]

These particular towed torpedoes were designed by Captain Menzing, a German, and closely resembled the British Harvey weapon, which will be examined in the next chapter. The Menzing was detonated by electricity with the aid of a firing key, as mentioned earlier, and it would seem its failures were almost exclusively due to this particular device as the weapon itself reached and struck the target on several occasions.[23]

Russia's first attack using the spar torpedo took place at Brailov on the river Danube on the night of 25–26 May 1877. Four launches under the command of Lieutenant Dubasov were employed and the spar weapons of the *Tzarevitch* and *Xenia* exploded under the keel of the Turkish monitor *Seifi,* tearing out her bottom and ensuring her ultimate fate. All four Russian boats that took part in the attack were badly damaged in the action but the success of the spar torpedo stood in sharp contrast to the lackluster performance of Captain Menzing's ineffective weapon.[24]

A Russian spar torpedo launch steaming into action during the Russo-Turkish War of 1887–88.

The next two spar torpedo attacks were, however, to cast doubts on the device's true worth. The first operation began when the steamer *Veliki Kniaz Constantine* sailed from Odessa on 9 June with six torpedo launches in tow. The initial target was the island of Fedoposi, which Turkish ironclads were known to patrol on a regular basis. But finding the area empty of warships, *Constantine* headed for Sulina, and some five miles from the anchorage her torpedo boats were dispatched on their le-

thal errand. Arriving at the entrance to the harbor at 2:00 A.M., they found three Turkish ironclads snugly at anchor while a fourth was patrolling the approach to the roadstead.

Lieutenant Puschschine and Lt. Zinovi Petrovitch Rohestvensky attacked immediately, Puschschine penetrating deep inside the anchorage and Rohestvensky steering for the nearest monitor. The third launch, under the command of Lieutenant Zazarenni, unfortunately tangled the wire used to lower the spar into the water in its own propeller and was forced to withdraw at slow speed.[25]

When Rohestvensky was only seventy feet from the *Idjilalie* Turkish lookouts spotted the Russian boat and shouted a challenge. And when there was no response, they opened a ferocious fire on the approaching launch, although unaware of either its nationality or its purpose. Braving the hail of bullets, Rohestvensky hurled his boat at the enemy, but, lowering the spar prematurely—or, according to Russian accounts, striking it against some form of boom or net defense—the warhead exploded without harming the ironclad monitor. The launch was less lucky. It was almost swamped by the blast from the explosive charge and, to make matters worse, the steering gear chose this critical moment to misbehave. Soaked to the skin and under heavy fire from the Turks, who were enjoying their enemy's discomfiture, Rohestvensky repaired the steering apparatus while his men cleared the wires that secured the smoldering warhead to the seaward end of the spar. Then, turning away, he used his speed to run clear of the rifle fire from the monitor's deck and courageously set off in search of Puschschine's boat—having guessed that a heavy explosion he had heard some minutes earlier marked a torpedo attack on one of the enemy ships anchored closer inshore.

But pursued by the *Idjilalie* against whom he could offer no defense, Rohestvensky was forced to break off his search and make his way back to Capt. Stepan Makarov, the Black Sea Flotilla commander, on his flagship *Constantine*. When steam launch *No. 2* returned to Odessa and was hoisted out of the water she was found to have a badly damaged keel where Rohestvensky had grounded on some form of static underwater defenses. He had, in fact, been doubly lucky for it was probably only the bow-up attitude of the launch, caused by the flooded stern section, that had allowed the lieutenant to recross the obstacle without further damage on the return run. Sadly, the gallant Puschschine's boat was never

seen again and was presumed lost in action with her entire crew. Captain Makarov, the flotilla leader, later achieved flag rank and died when his flagship *Petropavlosk* was sunk by a mine in March 1904 during the Russo-Japanese War. Zinovi Petrovitch Rohestvensky, the young skipper of steam launch *No. 2,* was to become an international figure some twenty-seven years later, when he led Russia's Second Pacific Squadron halfway around the world in a bid to lift the siege of Port Arthur. But he was intercepted by Adm. Heihachiro Togo off the island of Tsu-Shima on 27 May 1905, and his fleet, now reinforced by Admiral Nebogatov's squadron of old and barely seaworthy ships, was literally annihilated in the course of the subsequent battle. Rohestvensky survived, though seriously wounded, and became a Japanese prisoner-of-war. He died in retirement on 14 January 1909.

The Russian navy's next attack on the Turks was considerably more ambitious than the assault on the ships anchored off Sulina. This time it was carried out on 23 June in broad daylight at the mouth of the Aluta River. And the chosen target was not only moving, it also had its protective booms and steel nets extended. Only two launches were involved but, without the darkness of night to hide them from view, they were quickly sighted by the enemy gunners who drove them off with ease. Thus ended the spar torpedo's part in the conflict. Three attacks, one Turkish ship sunk, and one Russian torpedo launch missing. Although a better result than the towed torpedo, it was still less than impressive.

To complete this comparative review of the torpedo types employed during the Russo-Turkish War, it is necessary to consider briefly the part played by the Whitehead weapon during subsequent operations. On 27 December 1877, Captain Makarov used the *Kniaz Constantine* to tow two sloops, *Sinope* and *Tchesma,* to the roadstead at Batum for a night attack on the Turkish anchorage. Each carried a 15-inch Whitehead torpedo, *Sinope*'s being inside a discharge tube lashed to a wooden raft which was secured alongside, while *Tchesma* had her weapon slung beneath her keel in a rope cradle. Neither method was suitable for accurate launching.

Dashing recklessly toward the ironclad *Mahmoudieh,* the little *Tchesma* loosed off her Whitehead from a distance of sixty yards, but even at this point-blank range the torpedo missed and exploded harmlessly astern of the target. The *Sinope* followed suit, but her aim was even wilder,

and the Whitehead ended up stranded on a nearby beach. Although the method of launching was more at fault than the torpedoes, the attack, nonetheless, was a total failure. The aftermath of the operation revealed a strange story, however.

Henry Woods Pasha, a British officer serving with the Ottoman navy, claimed in his autobiography that, although one torpedo had exploded harmlessly, the Turks had captured two Whitehead weapons—one complete and one with its bow compartment missing.[26] He then alleged that, with an English friend, he had stripped down the intact weapon to reveal the Secret, namely, the balance chamber with its pendulum and hydrostatic valve mechanism that made the Flume-built torpedo superior to all of its rivals.[27] And it was the cost of the license to use this particular technology that had brought Robert Whitehead his fortune. On learning of Woods's discovery, the factory sent a special mission posthaste to Constantinople, headed, it is believed, by Whitehead's partner and son-in-law, Count Georg Hoyos, who was perturbed to find that the Turks now possessed the company's most valuable asset. And they had paid neither a penny nor a drachma for it.

According to Woods, it was agreed that both torpedoes would be shipped back to Fiume, where they would be repaired and then restored to the Ottoman government along with a further "three others of the most modern type" free of charge as a goodwill gesture. Finally a contract was negotiated under which fifty others of the latest improved pattern would be supplied to Turkey at half the price usually paid by other countries. It seemed an extraordinarily generous offer under the circumstances.

Part of Woods's account can be verified from a report in the *Times,* which revealed the serial numbers of the two captured weapons as 250 and 255, each with an additional marking of 30.[28] The Fiume factory records confirm these numbers as being from the first batch of thirty 15-inch torpedoes delivered to Russia in February 1877.[29] The same ledger revealed a mysterious entry for which no precise date was given. This showed a 15-inch torpedo as delivered to Turkey probably during 1878 and for which, significantly, no serial number was noted—the only occasion when this identification was omitted from the ledger in a period of forty years during which no fewer than 14,019 weapons were delivered.

It must also be assumed that the second torpedo which the Turks claimed to have captured—and which was confirmed in the *Times*—was

the one that was heard to explode but obviously did not. The fact that this particular weapon ended up on the beach "with its bow compartment" missing suggests that the detonator worked but failed to trigger the warhead—or there was some other similar malfunction that resulted in the warhead falling off. As it was not shown in the factory ledger as repaired and returned, it must be further assumed that it was too badly damaged to be restored to working order.

There are some other puzzles relating to Turkey's Whitehead torpedoes but as the factory's books do not seem to have been particularly well kept during 1877 and 1878 it is not possible to either specifically identify what happened or to solve the apparent mysteries that follow. In any event the Whitehead torpedo is another story and outside the scope of this book. Suffice it to say that the fifty half-price weapons promised by Hoyos do not show up in the delivery ledgers, the biggest consignment being for thirty-six torpedoes in June and July 1886 nearly ten years after these events. Perhaps, as General Berdan discovered in 1882, the Turkish treasury did not have enough cash to buy the full fifty weapons, even at half price.

The final attack with Whitehead torpedoes proved a resounding success and was again carried out by *Tchesma* and *Sinope* who, this time, actually got inside the harbor at Batum during a night operation on 25 January 1878, just six days before the armistice was signed. Both weapons were launched at a range of eighty yards and struck the revenue steamer *Intikbah* with an explosive force so great that their victim sank in less than two minutes with the loss of twenty-three men.[30] Hobart Pasha consistently denied the Russian claims but there is little reason to doubt their version of the facts.[31] A Reuters Agency report subsequently stated that "Captain Makarov announces that these torpedoes shattered the vessel, causing her to sink almost immediately. The broken spars and debris were seen floating on the surface of the water [and] the cutters were unable to save the Turkish crews."[32]

After the disappointing performance of the spar torpedo during the Russo-Turkish War, it is surprising to find that inventors continued to show an interest in this now outmoded weapon. Yet in January 1878, even before the conflict was over, William H. Mallory of Bridgeport, Connecticut, was granted a patent for his strange bifurcated spar, which, at its in-

Filled with Cork and Glue · 125

board end, was shaped like a wishbone.[33] The idea was to raise the spar in the manner of a tossed oar during the approach to a target and then, by means of a hand-operated worm and pinion mechanism, lower it down into the water until the charge was beneath the surface in the final moments before impact. The warhead was wired for electrical detonation, and Mallory suggested a double-ended boat which would have a similar spar at both bow and stern rather like Oliver Smith's "Armour clad" of 1862. The advantages to be obtained by lifting the charge out of the water during the attack run were not explained and the design does not seem to have progressed beyond the drawing board.

In 1880, Mallory came up with a more sensible, one might say novel, idea.[34] This was for a short-range free-running weapon which was intended to replace the spar torpedo launch, a pinnace often carried by battleships and classified as second-class torpedo boats. The accent was on simplicity and cheapness and the motive power was derived from a series of coiled springs that rotated the torpedo around its own axis "like a rifle bullet" to ensure accuracy of direction. The body shell was described as cigar-shaped and between twelve and fifteen feet in length with a diameter of ten inches. The propeller was placed centrally and mounted on a ring in the middle of the weapon—or, as Mallory termed it, "the boat."

Although no speed was given, the range was admitted to be only "a short distance—say five to six hundred feet," and it was intended for it to be carried inside, and launched from, a hollow tube fitted to the deck of an ironclad. Mallory provided a long description of the spring propulsion system which, unfortunately, makes matters no clearer. Yet it was an ingenious idea, and it would have been interesting to assess its performance in practical tests. This was not to be, however, and the practicality or otherwise of his "torpedo boat" remains unresolved.

Mallory was certainly bitten by the torpedo bug, for he went on to design a rocket weapon in which the propellant charge was contained in "a number of independent cartridge holders." The propellant itself was compressed gunpowder, but the inventor accepted that "other slow-burning gas producing bodies might be used." The device also incorporated a "novel method of compensating for the loss of weight . . . due to the destruction of the (gun) powder." This consisted of a collapsible diaphragm

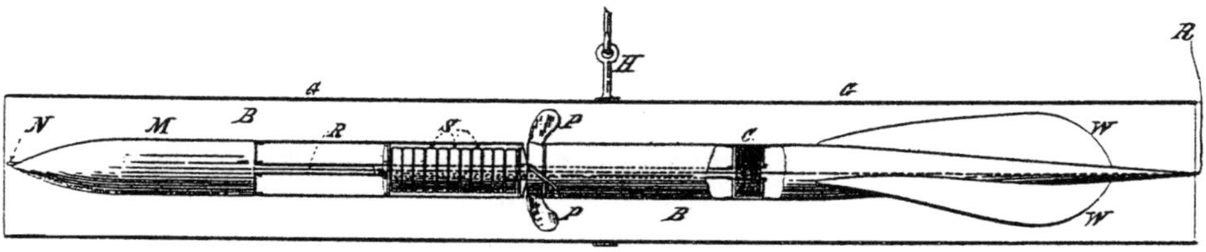

126 · *Nineteenth-Century Torpedoes and Their Inventors*

Designed by William Mallory and patented in 1880, this close-action torpedo had a diameter of only ten inches and a length of twelve feet. Propelled by a coiled spring, it rotated "like a rifle bullet" and had a maximum range of about six hundred feet—or so its inventor claimed.

provided with a "time-orifice" so that the specific gravity of the torpedo remained unaltered despite the loss of gunpowder due to combustion.[35] This, again, was purely a drawing-board project and was taken no further.

But let us retrace our steps. Two and a half months after Mallory's bifurcated spar was patented, Hans Mortensen of Alma, Colorado, followed suit with a new torpedo boat which was "capable of operating wholly underwater."[36] The specification, despite the accompanying drawings, was studiously vague, and it was clear that Mortensen had not really given his idea adequate thought. One thing was certain: the "submarine" would have been a deathtrap. For this reason alone the entire contraption does not merit further consideration.

Versions of the spar torpedo were still being dreamed up by inventors as late as 1889. And not by some crazy backwoodsman out of touch with reality but by a respected inventor and a pioneer of the modern machine gun, Richard Jordan Gatling. Born on 12 September 1818, Gatling was only twenty-one years old when he perfected a practical screw propeller, although in the patents race, he was beaten to the post by both Smith and Ericsson. However, like many other torpedo pioneers, Gatling's inventions covered a variety of industrial applications including, in 1850, a hemp-breaking machine and, seven years later, a steam plow. And although most famous for his machine gun he was, in fact, one of the innovators of mechanized farming in the United States.

The Civil War had brought him into contact with military armaments, and his first experimental machine gun, a single-barrel weapon with a rotary chamber produced in 1862 was subsequently developed into the

Filled with Cork and Glue · 127

famous Gatling gun, a multibarrel weapon that rotated about a central shaft by means of a hand crank and fired self-contained metal cartridges.

Gatling was a septuagenarian when he took out his patent for a "combined torpedo and gunboat" in 1889 and, even at that date it looked anachronistic not to say a little bizarre.[37] Gatling's intention was to construct a vessel that, by virtue of its design, protected its crew from hostile fire. It was fitted with a large gun in the bows and, below this, a spar torpedo on a special mounting that was "capable of a change of position"; that is to say, it could be pivoted up or down and moved laterally from side to side within certain fixed parameters. The unique feature of the patent was a small "annex" at the stern of the gunboat which contained the four-man crew that steered and otherwise controlled the vessel. They could enter or leave the main portion of the boat by a passageway at the stern, and Gatling thoughtfully added, "The operators may wear life-preservers." The spar weapon itself was not particularly novel and was only incidental to the general design. It was scarcely an original idea and was of doubtful practicality. Even worse, it was some twenty-five years too late.

It is encouraging to close this account of the much-maligned spar torpedo on an upbeat note, if only as a belated tribute to the gallant men who took it into battle. Despite its earlier disappointing showing, it saw further action on several occasions during the Franco-Chinese War of 1884–85—an old-style colonial struggle for control of Cambodia—where it covered itself in glory with a record of success that would have aroused envy even in the breast of its Whitehead cousin.

A daylight action off Foochow on 23 August 1884 resulted in a French victory inside seven minutes—certainly one of the shortest sea battles on record. Engaging what was, on paper, a superior force, Rear Admiral Courbet's squadron attacked without hesitation and reaped a deserved reward for their temerity. Torpedo boat *No. 46* sank the Chinese cruiser *Yang Wu* with an electrically detonated spar torpedo, while its flotilla mate, *No. 45*, damaged the sloop *Fu Sing* with a similar weapon. Later the same day, the crippled vessel was cornered and dispatched to the bottom by gunfire and a determined spar torpedo attack by an armed pinnace from Courbet's flagship.

By way of retaliation, Chinese launches equipped with spar weapons attempted a counterattack the following night but fled in confusion when

they found themselves exposed in the glare of French searchlights, the first time this new means of night defense had been tested in combat.

Six months later, on 14 February 1885, the French carried out a further night attack off Shaipoo when two torpedo boats armed with spar weapons sank the Chinese cruiser *Yu Yen* and a dispatch boat, *Chien Chiang.* It had been a brilliant little campaign with four Chinese vessels sunk in exchange for loss of one French sailor and minor damage to two of Courbet's torpedo boats.

But the spar torpedo's swan song was delayed for another ten years, and it claimed its last victim during the Sino-Japanese War of 1894–95 in the aftermath of the battle of the Yalu River in which a Japanese squadron had routed Admiral Ting's stronger fleet and forced him to flee to Port Arthur with his two battleships, *Chen-Yuen* and *Tin-Yuen.* During mopping-up operations on the day following the battle, the Japanese caught up with the cruiser *Yang Wei* and sank her with a spar torpedo.[38]

These final successes belatedly justified the simple concept of the short-lived spar weapon. Despite these last flickering flames of glory, however, it was no longer a match for the superior performance (and ever increasing complexity) of the all-conquering Whitehead. Bowing out gracefully, the spar torpedo quietly passed into retirement like a faithful old warhorse.[39]

CHAPTER 7

A Perfect Nightmare

L IKE MOST weapons in the field of underwater warfare, the towed torpedo originated in the fertile mind of Robert Fulton (1765–1815),
who, on 7 November 1800, told Gaspard Monge and Pierre-Simon de Laplace–the two scientists Napoleon Bonaparte appointed to report on the
American inventor's submarine *Nautilus*–that "there has come to me a
crowd of ideas infinitely more simple than the means I have employed
hitherto."[1] One of these ideas was the use of "carcasses of [gun]powder or
torpedoes" which were to be towed into a crowded harbor by his other
invention, the submarine. It was, in fact, only a second choice option.
Fulton's preferred method of attack involved thrusting a spike into the
enemy's keel and then attaching the nineteenth-century equivalent of a
limpet mine to it–in essence, a very close copy of the system employed
by David Bushnell with the *Turtle*.[2] But Fulton's device was proving obstinately reluctant to perform as planned because he was experiencing
problems controlling the submarine underwater.[3]

Although the concept of the towed torpedo did not occur to him until
1800, many writers who have studied his preliminary 1798 drawings of
the submarine[4] have mistakenly argued that this prototype of the *Nautilus* was intended to drag an explosive charge behind its stern in the
manner of a towed torpedo. The error is understandable as the drawing
included an explosive charge (labeled a torpedo in the key) which was
apparently attached to a line at the stern of the submarine. The sketch
also depicts "the horn of *Nautilus*," the fanciful name Fulton gave to the
spike he intended to be driven into the enemy's bottom planking–this being the device to which the explosive charge was secured during the final stage of the attack. The charge itself was detonated by a clockwork
time-delay mechanism.

[129]

On examining Fulton's drawing, the explosive charge to be used in conjunction with the spike is nowhere to be seen, which means that it must have been the "torpedo" at the stern. If such was the case, the towing line was, in fact, no more than a means of keeping the charge tethered to the submarine until the time came to secure it to the spike—or the horn of *Nautilus*. The single explosive charge illustrated in the drawing really allows no other conclusion. When Fulton originally sketched the submarine in 1798, he was thinking no further than a form of limpet mine, no doubt inspired by Bushnell. The concept of the towed torpedo entered his mind later as confirmed by his report to Monge and de Laplace.

In my previous book on the Robert Whitehead, I provided a concise summary of Fulton's life and inventions, and I do not wish to bore the reader by reiterating these facts.[5] However, as his work with towed torpedoes, brief though it was in terms of time, needs to be examined in some detail to prove that he was, indeed, the true creator of the weapon, it will be necessary to again cover some of that ground. This, happily, also affords me an opportunity to amend and correct an error made in *The Devil's Device* when I wrote that "according to a detail sketch made by Fulton in 1798 . . . [the *Nautilus*'s weapon] consisted of a single towed torpedo—the first such device to be built and a direct ancestor of the infamous Harvey torpedo."[6] As already noted, this statement is incorrect. For as we will see, the towed torpedo, when it finally appeared, had no connection with his submarine of which, by that time, Fulton had already tired.

Although he continued his experiments with *Nautilus,* which by 1801 could remain submerged for up to three hours with four men aboard, Fulton was developing his "submarine bombs."[7] These were copper containers holding up to two hundred pounds of gunpowder which could be detonated on contact with the target by means of a gunlock. Despite claims that he used the submarine for his first public demonstration of the new weapons which he had organized for the benefit of Admiral Villaret, modern historians are generally agreed that Fulton employed a pinnace crewed by twenty-four seamen who operated four manual cranks to propel the boat through the water.

A forty-foot sloop was chosen as the target, and on 12 August 1801 the American inventor steered the pinnace, towing one of his smaller twenty-pound "submarine bombs," toward the anchored target. When he had

closed to within twenty yards of his prey, Fulton altered course sharply and sent the "torpedo" curving in a wide arc toward the sloop. There was a loud explosion, and as the smoke and spray cleared, the stricken vessel could be seen settling in the water. Moments later it vanished beneath the surface as the circling seagulls screamed a shrill farewell. It was the towed torpedo's debut. And it had immediately claimed a victim. There could be no mistaking the portents of Fulton's success.[8] The details of this demonstration are supported by considerable documentary evidence which prove, beyond doubt, that Robert Fulton invented and built the first practical towed torpedo.

It is therefore somewhat curious that he did not follow up his success and carry out further experiments with the weapon. For when he became involved with the Royal Navy in its attack on the French invasion fleet at Boulogne on 2 October 1804[9]–he had by this time changed sides after failing to obtain what he wanted from the French government, which seemed to be losing interest in his inventions–the primary weapon he proposed to employ was the drifting mine. Various other devices were also used, notably a Tudor-style fireship packed with forty barrels of gunpowder under a layer of flint chippings, a combination almost as lethal as a modern nail bomb. This twenty-one-foot vessel weighed around two tons and detonation was controlled by one of Fulton's favorite devices, the clockwork time-delay mechanism which he had developed for the *Nautilus.* This centerpiece was supported by fifteen similar, but smaller, "bomb" vessels plus a number of oared boats towing large carcasses of explosives. The entire force was under the command of Admiral Lord Keith and there was considerable alarm among the French when the flotilla arrived offshore during the early evening led by the seventy-four-gun *Monarch* together with frigates, sloops, and various small gunboats.

When the admiral's flag hoist signaled the start of the attack, the safety pins were withdrawn from the gunpowder carcasses, or torpedoes as they were also known, which were then cut free and allowed to drift on the incoming tide toward the serried ranks of moored shipping. As an additional threat to the invasion fleet they were accompanied by the unmanned and similarly drifting bomb vessels and fireships. The night sky was lit by a series of brilliant flashes while the roar of cannons and the rattle of muskets echoed across the black waters. The whole affair was more spectacular than effective, however, and the French ships and

132 · *Nineteenth-Century Torpedoes and Their Inventors*

barges suffered little damage. It needs to be emphasized at this juncture that, although the "torpedoes" were towed into action, they were cut free and left to their own devices when the attack began, relying solely on the wind and tide as their source of propulsion. They were thus drifting mines, not towed torpedoes.

Fulton's next weapon, which was demonstrated near Dover on 15 October 1805, just days before the battle of Trafalgar, was another development of the drifting mine concept.[10] This time two explosive canisters, linked together by a line, were carried to the target, the brig *Dorothea*, by a pair of small rowing boats. The line was grappled to the vessel's anchor chain and the clockwork delay mechanism was set. Precisely fifteen minutes later, as Fulton had confidently predicted, an almighty explosion rocked the sturdy walls of Prime Minister William Pitt's country home, Walmer Castle, and, torn apart, the *Dorothea* vanished beneath the surface of the English Channel in, to quote Fulton's own words, "exactly twenty seconds" to secure her unenviable niche in history as the first large boat to be sunk by a mine. Once more, it should be noted, the explosive device was most certainly not a towed torpedo and is best described as a form of limpet mine lashed to an anchor cable.

The inventor's flintlock harpoon gun designed for use against moving ships proved to be neither fish, nor fowl, nor good red herring.[11] And, again, it was not a towed torpedo. A small bow-mounted musket was used to fire a harpoon or bolt into the hull of an enemy vessel and a sixty-foot line was attached to the eye of the harpoon with an explosive device secured to the other end. And inevitably a clockwork delay device was employed to enable the oared boat to escape out of range before the charge was detonated. It was a cumbersome and highly dangerous way in which to sink an enemy ship. In a letter to President Thomas Jefferson dated 28 July 1807, Fulton indicated that he had modified the weapon by connecting the torpedoes with "a line and chain 100 feet long"[12] similar to chain shot, although the link was a great deal longer. Supported by floats, and suspended some ten feet below the surface, the charges were then left to strike the enemy's hull by the unpredictable action of the wind, tide, or current. It was the Boulogne drifting mine all over again—only slightly different. But it was still not a towed torpedo. As Fulton must have given a great deal of thought to each of these new devices it is difficult to understand why, having dreamed up the towed torpedo in August 1801,

he made no attempt to resurrect and develop what was very clearly a viable weapon.

Fulton's explanations are frequently confusing and ambiguous. This latter defect, to judge by his overly optimistic sales patter in support of *Nautilus* and his personal arrogance together with a predilection for self-advertisement, being probably deliberate. And in this context, the illustrations in his 1810 pamphlet *Torpedo War and Submarine Explosions,* especially plate V, which depicts a multiboat attack on a ship taking evasive action, have misled many writers into believing they showed an attack with towed torpedoes.[15] At first sight this, indeed, appears to be the case. But careful examination suggests that we are looking at an attack by rowing boats armed with Fulton's harpoon gun. The two "torpedoes" shown as almost touching the target's hull are not linked by a common chain but appear to be separately attached to the enemy by means of a harpoon or bolt. In confirmation the shaft of the harpoon spear can be seen protruding from the target's bows. The small stub which projects from the bows of each oared boat might depict the barrel of a harpoon gun.

Fulton had by now seemingly abandoned the towed torpedo despite its promising debut, although he occasionally referred vaguely to such weapons. Instead, he turned his attention, as John Ericsson was to do some sixty years later, to a "submarine gun," or *Columbiad,* as he christened it. This weapon, dating to 1813, consisted of two smoothbore cannons which were to be fired submerged so that "the bullets will pass through the water instead of through air, and through the side of the enemy's vessel below the surface which, letting in the water, will sink the vessel."[14] He neglected to explain how the fuse was to be kept burning underwater or how he proposed to solve the other obvious problems involved.

But within a few months, as we saw in the previous chapter, his agile mind had moved from the submarine gun to the spar torpedo. And the diagram he sent to his friend Capt. Stephen Decatur in September 1813 is proof that he was the originator of this new concept in naval armaments. Yet as had been the case with the towed torpedo, Fulton made no effort to develop the spar weapon, although the fact that he died on 24 February 1815, just eighteen months after communicating with Decatur, might account for this omission, as he was reported to be exhausted and overworked a short while before his death.

It is rare for a man to dream up two major weapons of war in a single lifetime. But Robert Fulton did so with the spar and towed torpedo even though he subsequently failed to exploit either idea. He had also designed and built one of the world's first practical submarines while, in a different field altogether, he pioneered steam navigation on America's East Coast river system. It was a remarkable achievement for a Pennsylvania farmer's son who, in his younger days, had earned his living painting, and selling, miniature portraits.

The towed torpedo, once described by Bethell as a "perfect nightmare" and "this deplorable weapon," was resurrected by two Royal Navy officers, Capt. John Harvey and Cdr. Frederick Harvey, during the 1870s.[15] Contrary to popular opinion they were not brothers but uncle and nephew and their relationship was confirmed by a letter which John wrote on 29 May 1871. In this, he referred to Cdr. Frederick Harvey as "my nephew" and explained that "the problem of the sea torpedo . . . has engaged my thoughts over a period of a quarter of a century." He continued, "I have, however, long since relinquished attempts to solve it, my nephew having, for some years, devoted his attention to the subject."[16]

From this we can conclude that John Harvey, who by May 1871 was signing himself as "retired Captain RN," first began thinking about the towed torpedo before 1850. Much earlier than is often supposed. It would also seem that his pioneering experiments were taken over by his nephew, Frederick, and that it was the latter who was responsible for the towed torpedo as it finally emerged shortly before 1870. However, as will be seen later, there was often considerable confusion in both contemporary newspapers and the technical press as to which was which, and the English aversion to providing the Christian names of officers only served to compound the difficulties of researchers.

By employing a form of "otter board" of the type used in deep-sea fishing to hold the mouth of the trawl net open and fastening to it an explosive charge, they found that they could "swim" their torpedo at 45 degrees from the towing vessel's longitudinal axis once it had reached a critical speed of around six knots. In the course of exhaustive experiments it was found that the optimum length of the tow line was, in practice, about 150 yards. It is interesting to note that John Harvey referred to the weapon as the Harvey *otter* torpedo on several occasions, although this descriptive name failed to stick and the Harvey towed torpedo

seemed to be preferred by both press and public. It was, of course, simple and it was cheap. And that was the way Her Majesty's Treasury liked things. Why, then, was the Royal Navy so reluctant to adopt it as part of Britain's naval armory?

It is apparent from subsequent comments by John Harvey that he and his nephew had made a number of approaches to the Admiralty—none of which aroused any interest in the right quarters. And the first mention of the weapon so far traced appeared in a report in a professional journal in October 1869:

HARVEY'S SEA TORPEDOES

A considerable number of these formidable machines have been supplied by the maker, Mr William Nunn, of St George Street, London Docks, to the Russian government.[17]

A further report in March 1870 helps to fill in the gaps.[18] It is worth observing at this point that while many writers tend to treat the Harvey as a predecessor of the locomotive torpedo it was, in fact, an exact contemporary. And this probably accounts for the reluctance of the Admiralty to adopt it. The Torpedo Committee, which examined the Harvey along with the outrigger or spar weapon together with the Whitehead, Lay, and von Scheliha torpedoes, showed a similar lack of enthusiasm.[19] And apart from its continually improving performance, the Whitehead was an altogether less exasperating—and, indeed, safer—weapon to operate under combat conditions. But to return to the report in *Engineering*.

The torpedo was described as having a stout wooden casing, strengthened on the outside with iron plating, and incorporating a metal shell which contained seventy pounds of a mixture known as Horsley's powder, claimed to be fifteen times more powerful than the best gunpowder. The report actually provided details of the ingredients and proportions required to make Mr Horsley's concoction which in today's terrorist environment, the author considers it wiser to omit. One significant aspect of the Harvey which has not been found in either contemporary papers or more modern publications was revealed in a single short sentence: "The ingredients are kept separate and are mixed in a sieve *at the time of use*."[20] This tiny detail goes a long way to explain why the admirals lacked enthusiasm for the Harvey while the armchair pundits of the technical press, not being at the sharp end, were all in favor of the device. Can

the reader imagine mixing two ingredients of an explosive claimed to be fifteen times more powerful than gunpowder, in a sieve on an open deck, and then pouring the results into a container comprising the torpedo's charge before it was streamed alongside in readiness for an attack? And all of this carried out under heavy fire with shells bursting in all directions and the torpedo boat being raked by rifle and Gatling-gun bullets? The professional seaman's reluctance to become involved in such suicidal behavior is understandable.

There is, however, a sound reason for supposing that this section of the press report was misleadingly fanciful. In his *Instructions for the Management of Harvey's Sea Torpedo,* a copy of which only came into my hands after the chapter was completed, Cdr. Frederick Harvey revealed that, in fact, eight different explosives were available and Horsley's powder was but one of them. He listed their names and weights. (See table below.)

He offered no comment on the employment of the charges or their relative merits but added, "Some of the powders named have not yet been manufactured on a large scale."[21]

It is apparent from the *Instructions* that the detonating material was supplied inside special containers known as "explosive bolts" which were inserted into the torpedo while it was being prepared for launching and there is reference to a "bolt magazine" inside which these devices were stored on board ship. Harvey added in a footnote, "As the certainty of explosion depends upon the exploding bolt being properly charged, the

Explosive	Large torpedo (lbs.)	Small torpedo (lbs.)
Glyoxilin	47	16
Compressed guncotton	60	22
Schultze's blasting powder	60	22
Picric powders	73	26
Rifle grained powder	76	27
Horsley's original	80	28
Horsley's blasting powder	85	30
Nobel's dynamite	100	35

inventor [i.e., Harvey himself] takes entire charge of this important detail."[22] As the commander could not be in more than one place or one ship at a time it is clear that the bolts were precharged ashore and stored in some form of magazine until required for use.

In light of this footnote it must be assumed that the journalist who wrote the report in *Engineering* had been the victim of an overimaginative sailor with a misplaced sense of humor for there is no evidence whatsoever in the *Instructions* that any of the explosive powders were mixed together on board. Indeed Harvey's footnote makes it clear that all of the detonating material was prepared and prepacked ashore.

The otter board to which the explosive charge was secured was designed to diverge some 45 degrees from the fore-and-aft axis of the towing vessel and "an arrangement of slings in connection with it, enable[s] the operator to diverge the shell alongside the enemy ship in meeting, passing or crossing, whichever method of attack is adopted."[23] Reference was also made to the attacking vessel's "great speed." But as already noted, experience had shown that ten knots had proved to be the optimum rate of towing. We learn, too, that the running depth of the weapon could be regulated by the towing speed, while a buoy attached to the casing prevented the torpedo from submerging too deeply. The explosive itself was detonated when an external lever came into contact with the target and actuated an internal primer. The dimensions of the otter board were given as length, four feet six inches; depth, two feet; and width, six inches—and two different explosive charges could be fitted, one consisting of seventy-six pounds of gunpowder and the other one hundred pounds of dynamite. This particular report made no mention of Horsley's powder which may, by then, have been abandoned in favor of either gunpowder or dynamite.

Indeed by this time great emphasis was placed on the safety key which "relieves the operators from all fear of accidental explosions [for] until the key is withdrawn the exploding apparatus will not act, and the key is never withdrawn until the shell [*sic*] is some yards astern of the torpedo vessel." It will be recalled that Fulton had fitted some form of similar device for his bomb ships at Boulogne in 1804. There seemed to be very little new about the towed torpedo.

The purpose of this particular report was to provide details of the Admiralty's trials of the Harvey "recently carried out"[24] off Spithead, Ports-

mouth, under the supervision of Capt. Henry Boys, the commanding officer of HMS *Excellent*, the Royal Navy's gunnery establishment which, in 1870, was also responsible for torpedoes, sea mines, and, somewhat surprisingly, electrical devices and apparatus.

According to *Engineering* ten simulated attacks were carried out on the obsolete turret ship *Royal Sovereign* by the steam tug *Camel* towing a single seventy-six-pound torpedo fitted with an explosive bolt in lieu of a detonator. The towline was fifty fathoms (three hundred feet) in length and the attack speed was recorded as eight knots. The results were impressively good and *Camel*'s torpedo obtained a contact on every single run. But as the target was anchored, success was relatively easy. Each torpedo struck *Royal Sovereign* below the waterline—the depth varying between one and sixteen feet—while the explosive bolts provided proof positive that a hit had been secured on every occasion. The runs had been made from ahead and astern and also while crossing the target's bows. During each of the simulated attacks, *Royal Sovereign*, using blank charges, had fired her turret guns to measure the degree of defense that an enemy ironclad could employ during an attack with a towed torpedo. On eight of the runs she only had time to loose off two shots, although on the remaining two runs her gunnery officer managed to fire four and seven blanks, respectively—an improvement which suggested that practical experience was an important factor in the potential enemy's response. Finally it was noted that the safety key had been successfully withdrawn "at distances ranging from 8 to 60 yards."

The second series of tests was more ambitious and realistic. *Royal Sovereign*, a converted 131-gun line-of-battle ship, was proceeding under weigh at between eight and nine knots while *Camel* was steaming two knots faster. This time, and assuming that the report was accurate, the tug was streaming two torpedoes—one from each of her bow quarters—with a towline length of, again, three hundred feet. Captain Boys, who was in command of the turret ship during the exercises, tried to steer *Royal Sovereign*, once described as "the ugliest battleship ever built," out of range, but all attempts to escape failed, and in the six runs carried out, it was stated that "*every* torpedo invariably struck the adversary."[25] To achieve such a result meant that *Camel* had made two passes during each attack so that both the port and starboard torpedoes were aligned relative to the target. This point, however, is not mentioned in report. Similar tac-

tics were adopted as those employed for the previous set of exercises against a moored target—from ahead, astern, and crossing—and again the torpedoes struck the turret ship's hull from one to sixteen feet below the waterline. In fact, two actually hit *Royal Sovereign*'s keel. For this exercise the ironclad succeeded in firing between two and twelve rounds from her five cumbersome 10.5-inch guns during each run.

Engineering's report was euphoric in its assessment of Captain Harvey's weapon. "The results of these trials proved the torpedoes to be perfectly under command and thoroughly effective in action. One great feature of the torpedo is that all its arrangements are simple and sailorlike, and it is exactly fitted for the class of men within whose province it will lie to use it [a back-handed compliment if ever there was one]. The paying-out apparatus and its brake arrangement are also readily worked by an ordinary seaman; in fact from first to last there is nothing that requires a specially experienced staff to work this torpedo."[26]

The writer also pointed out that if, at the last moment, the target was found to be friendly "her destruction can be arrested. This [is] effected by so managing the tow-rope . . . that the torpedo [avoids] the approaching vessel." A delightfully vague observation clearly intended as a dig at the fish torpedo, which, once launched, could be neither stopped nor diverted—this latter characteristic, it will be recalled from chapter 1, having always been a source of worry to Victorian pundits who, it would appear, remained blissfully ignorant of the fact that a 12-inch shell from one of their beloved "big" guns was equally difficult to stop once it had left the barrel.

Finally, the periodical urged all concerned to adopt the Harvey torpedo without delay. "The Government having proved its merits will, we presume, recognise them by adopting it into the [Royal Navy] without further delay." And it must be admitted that it was, indeed, a glowing report of a demonstration that seemed to yield near-perfect results. But the Admiralty did not respond by placing an immediate order, despite pressure from certain quarters within its own ranks.

It was also noted in chapter 1 that one of those who favored the towed torpedo was John "Jacky" Fisher, who, although only promoted to the rank of commander in June 1869, was already recognized as one of the brightest brains among middle-ranking officers in the Royal Navy, a reputation based largely on his own pushing methods of self-advertisement.

For following his visit to the Kiel mining trials that year, he continually drew attention to himself as the navy's torpedo expert. He was, in fact, in charge of all torpedo instruction at HMS *Excellent* and thus had some justification for his conceit. But for a man of his stature, professional knowledge, and grasp of modern naval weapons and tactics, it is surprising to find him wholly in favor of the towed torpedo and scornfully opposed to the Whitehead self-propelled weapon despite confidential inside knowledge of the latter. His view that the Harvey torpedo was "far superior" to the fish weapon has already been mentioned and was confirmed after he left *Excellent* on appointment to *Donegal* for a tour of duty on the China station on 25 November 1869. His departure from England thus took place before the Fiume fish torpedo was officially tested by the Royal Navy in the Thames estuary in the autumn of 1870 and contracts were exchanged between the British government and Robert Whitehead in April 1871.[27]

For some strange reason no one bothered to keep Fisher appraised of developments in England—or if they did, he never mentioned the fact—even though one of his regular correspondents was Capt. Henry Boys, who, as noted, took a leading part in the Harvey torpedo trials in March 1870 and was still in command of *Excellent* in 1871, when work was about to commence on the construction of the Whitehead under license in the Royal Laboratory workshops. The first Woolwich-built torpedo was, in fact, tested in May 1872,[28] almost exactly a month before Fisher returned to England in *Ocean* before her decommissioning on 22 June 1872. And throughout his time in Chinese and Australian waters, Fisher continued to play with his version of the Harvey weapon which he was trying to adapt to electrical firing.[29] As will be seen later there could have been another reason for Fisher's inexplicable support for the towed torpedo and the Harvey in particular.

The Admiralty partially responded to this lobbying campaign by appointing "Captain Harvey" to instruct officers and men in the use of his weapon at the two gunnery establishments *Excellent* (Portsmouth) and *Cambridge* (Devonport).[30] In addition, he was asked to supply one large and two small torpedoes for use on *Excellent* "as well as other torpedoes for the Service." The journal added smugly, "We, moreover, learn with satisfaction that Captain Harvey has been invited—when the proper time

shall have arrived–to place before the Government his claim for reward on account of time spent and expenses incurred in bringing his invention to its present state of efficiency."

A report in *Engineering* in January of the following year mentioned for the first time the London Ordnance Company and alleged that it was manufacturing Harvey torpedoes "on a large scale" as a speculative venture. "Other powers, we believe, are following in the wake of Russia," it added ominously.[31] This was not entirely accurate. Germany, France, and the United States all designed and constructed their own towed torpedoes; they did not buy the Harvey weapon.[32] But in a final *Stop Press* paragraph the journal revealed that the Admiralty had now ordered twenty Harvey torpedoes and added, a trifle petulantly, "Twenty, however, is a small number, and the order is very inadequate to the full requirements of the service."

There was an odd item in *Engineering* the following month. Quoting from the *Broad Arrow*,[33] it revealed that "Commander Frederick Harvey has lately applied to the Admiralty for promotion, and has been informed that his invention gives [him] no claim beyond other officers."[34] It is the next sentence, however, that arouses curiosity, as it derives from *Engineering* itself and not the *Broad Arrow:* "Now the original inventor of the [towed] torpedo is Captain John Harvey RN, but his nephew, Commander–Captain by courtesy–Frederick Harvey, RN, has for years past . . . developed it to its present successful issue." The phrase "Captain by courtesy" suggested that many references to "Captain Harvey" in the press applied not to (Capt.) John Harvey, but to his nephew (Cdr.) Frederick Harvey. Once again the English aversion to the use of first names had created a researcher's nightmare.

Fortunately, the very next sentence confirmed the suspected confusion between the two officers and, happily, clarified their identities: "He [Frederick] it was who had carried out all the experiments with the torpedo, and *who has instructed the officers and men of the Royal Navy in the use and manipulation of the weapon.*" Thus the "Captain Harvey" appointed to *Excellent* in 1870 was, in fact, Cdr. Frederick Harvey, not his uncle. The report also revealed that "it was Captain John Harvey who applied for the promotion of his nephew, presuming that his ordinary course of service, being supplemented by that of the hard mental labor which for some

years he had bestowed upon the torpedo, gave him a well-founded claim for promotion." Their lordships did not agree and there seems little doubt that this press campaign for both the adoption of the torpedo and the promotion of Frederick Harvey did little to help the two hapless inventors who appeared to have become mere pawns in an orchestrated attack on the establishment. The final sentence of the report seems to bear out this conclusion: "We trust we shall yet hear that Commander Harvey's just claims to recognition have been admitted by my Lords at Whitehall."[35]

Forty years later one would have suspected the hand of Jacky Fisher behind this type of lobbying campaign, for he was a past master in the art of using the press to achieve his objectives. And, of course, not only did he favor the towed torpedo as a weapon, but he was probably already personally acquainted with Frederick Harvey from his days on the instructional staff of *Excellent* between 1866 and November 1869, even though the latter was not actually appointed to the establishment until the spring of 1870. Unfortunately, we will never know for certain what lay behind this agitation, for at that particular time Fisher was safely on the other side of the world.

On Saturday, 22 July 1871, a small paddle-wheel steamer, the *Andrew Wodehouse*, "attacked" ships in Yarmouth Roads, an anchorage on the northwestern coast of the Isle of Wight, with a single deactivated towed torpedo under the control of "Captain" (almost certainly Frederick) Harvey. The exercise seems to have been nothing more than a publicity stunt organized by the manufacturers, the London Ordnance Works, but it must have caused some alarm and despondency among the local mariners. According to the report,[36] "the first attack was made against a brig coming in under full sail. The brig yawed, but the torpedo was successfully dipped under her keel with about thirty fathoms of line on it. The second attack was against a brig at anchor . . . [and] . . . was again dipped successfully under her bottom, both the firing levers being driven home [and] piercing the detonating capsule." Further attacks were reported on other vessels both under sail and at anchor while the speed of the sixty-ton *Andrew Wodehouse* during its final run was stated as eleven knots.

This time the dimensions given for the "small" Harvey differed from those provided in the previous report published a few months earlier. For clarity, therefore, the definitive measurements given in Harvey's *Instructions* for the two weapons are as follows:

Exterior Case	Large Torpedo	Small Torpedo
Length	5 ft. 0 in.	3 ft. 8 in.
Breadth	0 ft. 6.125 in.	0 ft. 5 in.
Depth	1 ft. 8.75 in	1 ft. 6 in.

The exterior case was said to be constructed from "well-seasoned elm" with a thickness of one and a half inches, and the interior case was made from "stout sheet copper."[37]

Harvey also pointed out that both torpedoes were constructed in two versions "so that one may diverge to port, and the other to starboard." This divergence meant that the detonating contact levers were located on either the right or left hand side of the torpedo dependent upon whether it was a "port" or "starboard" weapon for, as Harvey observed, the torpedo "has the advantage of exploding only when in hugging contact with the vessel attacked."[38] Thus the charge was detonated not so much by impact as by a gentle caressing movement along the victim's beam, which depressed the firing levers into the primer. As a result, the levers had to be placed on the left-hand side of the starboard torpedo and on the right-hand side of the port weapon.

One serious flaw in the towed torpedo's design is highlighted on page 24 of the *Instructions*, where Harvey observes, "It would be better to stop the screw, if circumstances would allow of it, when lowering the torpedo and buoys into the water, to prevent the chance of the buoys fouling the screw." This latter problem was common to most cable-powered or cable-guided dirigible torpedoes but this is the only occasion when the danger has also been linked to the towed weapon. To bring the attacking vessel to a standstill in order to stream the otter boards in full view of the enemy suggests that, for all the brave words, the Harvey towed torpedo was just as suicidal under combat conditions as the spar weapon. And it was presumably recognition of this disadvantage that persuaded the Royal Navy to keep the device at arm's length for as long as possible.

The London Ordnance Works, which had by now taken over production of the Harvey weapon from Mr. Nunn, was reported to have built "a large number" of the torpedoes at its factory in the borough of South-

wark,[39] this establishment being subsequently traced to Bear Lane.[40] As Nunn's workshop was located in St. George's Street, London Docks, in what is now the adjacent borough of Rotherhithe, it can be assumed that the London Ordnance Works did not take over Nunn's premises but set up its own plant three miles farther to the west. St. George's Street has now vanished from the map, but St. George's Stairs are still marked on the southern bank of Limehouse Reach, where the Thames flows almost directly south before looping east and north around the Isle of Dogs.

When and why John and Frederick Harvey removed the manufacture of their torpedo from Nunn and transferred production to the London Ordnance Works remains a matter of conjecture but the changeover certainly occurred no later than 1871 for Frederick Harvey included a footnote in his 1871 *Instructions:* "The torpedoes here described are manufactured by J. Vavasseur & Co, at the London Ordnance Works; at which establishment the inventor has every facility for the supervision of the various details in the construction of the torpedoes, buoys, and brakes. Such supervision of the torpedoes and their equipment is highly essential to secure safety and efficiency."[41] The significance of J. Vavasseur and the London Ordnance Works will become apparent shortly.

The 22 July tests were repeated on Saturday, 25 November, and the chosen venue was, again, the Yarmouth Roads.[42] But this time the tug which had been hired to tow the two torpedoes had Russia's General Prestich and the military attaché from France's London embassy, Baron de Graney, on board to observe the demonstration. The first attack featured the "large" torpedo and was carried out from astern on a, presumably, unsuspecting brig proceeding under full sail. There were smiles all round when a hit was secured with a tow line length of 60 fathoms (360 feet). But a second attack, this time with the "small" weapon, proved less successful after the wire towing cable parted while it was passing under the target's keel. Harvey explained this away; it was, he said, the result of damage to the hawser during the earlier experiments in July. Following this mishap the demonstration was abandoned for want of suitable targets, but the two foreign observers expressed themselves as satisfied with what they had witnessed.

Engineering's report revealed that "the operations were carried out by Captain Harvey [on this occasion undoubtedly Frederick] assisted by Mr. J. Vavasseur, of the London Ordnance Works." It is the mention of Josiah

Vavasseur's name that brings Jacky Fisher back into the saga of the towed torpedo. Thirty years later, Vavasseur was a close friend of Fisher and his wife. He was also, from 1883, a director of the powerful armament's manufacturer, Sir W. G. Armstrong, Whitworth and Company. Vavasseur had founded the London Ordnance Works in 1860 and had specialized in the development of gun mountings, culminating in 1871 with a hydraulic recoil cylinder which formed the basis of both his reputation and his fortune. Vavasseur was, sadly, childless and with his wife "privately resolved to arrange for Fisher's son, Cecil, to benefit directly and thus for Fisher himself to benefit indirectly."[43] Fisher's biographers, if they have mentioned the arrangement at all, have remained discreetly quiet about the details, and as Vavasseur died on 13 November 1908, it can only be presumed that Cecil Fisher was a major beneficiary under the terms of the will. As a result Vavasseur's country estate of Kilverstone in Norfolk passed to Fisher's son.

The link between Vavasseur and the Harvey towed torpedo has not previously been recognized, but in view of Fisher's involvement in all forms of torpedo warfare in the early 1870s, it would be strange if the two men were not, at the very least, acquaintances. In addition, Fisher was the director of Naval Ordnance from October 1886 to May 1891 and, again, would have met up with him in the course of his duties, for both were closely involved with Naval Ordnance and Vavasseur's gun mountings had been purchased by the Admiralty in 1882, although this was before Fisher was in any position to influence policy.[44] Yet Vavasseur held Fisher in such esteem that he bequeathed his large country estate to the latter's son, carefully leapfrogging Fisher himself for obvious reasons. But why? Unfortunately, we do not know, but there are some vague clues.

It has always been a puzzle why Fisher, a sensible and intelligent man, threw all his weight behind the Harvey torpedo and, in its early days, led a resolute attack on the Whitehead weapon, although he was to perform a U-turn on the automobile torpedo within a few short years. In the light of *Engineering*'s report, it is now apparent that he probably knew Vavasseur as early as 1870 or 1871—hence his support for his friend's interests.

Vavasseur's decision to pass Kilverstone on to Cecil came at a time when Fisher was going through a difficult period in his career. Uncertain that he would be appointed as first sea lord, the vital summit of his career if he was to succeed in reforming the Royal Navy from top to bottom, he

was even considering the possibility of retirement. Fate intervened, however, and on Trafalgar Day 1903, he became the executive head of the Navy. In a letter dated 11 September 1904, written to Lord Esher, he wrote, "I asked for my pay [as first sea lord] . . . to be made up to £3,400 a year on account of my giving up an appointment of £4,000 a year to suit them."[45] Later, in March 1905, when he was again trying to obtain more money from the Treasury, "he dropped hints of further offers from private industry–about £20,000 a year."[46] Finally, on 3 December 1905, during an unannounced call at 10 Downing Street, he told Prime Minister Balfour's personal secretary that he had been offered a "commercial post" at ten thousand pounds a year and wanted to know whether Balfour thought he should take it.[47] Much of this could have been bluff–or even examples of Fisher's sometimes odd sense of humor. But there emerges an underlying impression that these offers had emanated from Armstrongs and Josiah Vavasseur. Certainly the reference to an appointment of four thousand pounds a year which he had given up and which he would have been holding while a serving flag officer poses questions that have never been either asked or answered.

One final, almost trivial, detail reveals more than Fisher may have realized. When he was raised to the peerage on 9 November 1909, almost exactly a year after Vavasseur's death, his understanding of the armament tycoon's intentions was made clear for all the world to see. For he took the style of Baron Fisher *of Kilverstone.*

By 1872, even royalty was showing an interest in the towed torpedo, and on Friday, 22 March, HRH Captain the Duke of Edinburgh, RN, embarked on board the Admiralty steam tug *Grinder* to observe exercises with the turret ship *Royal Sovereign* in the area between Spithead and the Nab lightship.[48] It was virtually a repeat of the tests held in March of the previous year, but on this occasion only four of the five attacks succeeded (the reason for the single failure was not stated). It was noted that "His Royal Highness landed from the *Grinder* at Southsea pier after the conclusion of the practice and, shortly afterward, left for London." Perhaps, like his mother, he was not amused.

Shortly after the royal visit, although entirely unconnected with it, the Admiralty revealed that it intended to withdraw the inventor's "control over the torpedo's manufacture and proposed to transfer all construction of the weapon from Bear Lane to the Royal Laboratory at Woolwich." This

would, of course, have resulted in a considerable loss of income for Vavasseur and Company, which now owned the London Ordnance Works. There was an immediate, and inevitable, outcry from the professional press, and, bowing to "Captain" Harvey's personal protests, the Navy placed an order for a further twenty-five weapons, a placatory gesture that effectively postponed the threatened move.

On learning of the reprieve, *Engineering* commented, "We trust this arrangement will be permitted to remain undisturbed, as in justice to the inventor [*sic*] it ought; as well as out of consideration for the manufacturers, Messrs Vavasseur & Co, who upon the faith of the original arrangement, laid down special plant and machinery for making the torpedoes."[49]

As a result of this unseemly wrangle, it emerged that Harvey–presumably Frederick, although this was not made clear in the report–had asked for a payment of £5000 (the equivalent of at least £200,000 in today's terms) for the time and money spent developing the weapon but was only given £1000 "with an intimation that he might make what he could out of the manufacture of the torpedo."[50] As the government, in the previous year, had agreed to pay Robert Whitehead £15,000 for the non-exclusive manufacturing rights of the fish torpedo plus the entire costs of the 1870 trials, it is clear that their lordships did not hold the Harvey towed torpedo in very high regard. Nevertheless, these continual claims and counterclaims between the Harveys and the Admiralty leads to a suspicion of murky dealings somewhere in Whitehall's corridors of power. And the proposal to transfer manufacture of the weapon from the London Ordnance Works to Woolwich suggests that the Treasury had an uneasy feeling that it was being overcharged by the inventors and their associates.

Despite this background of simmering antagonism, development of the Harvey torpedo continued. For example, a method of electrical detonation had been devised in or before 1871 by Capt. C. A. McEvoy,[51] although for some reason this was not demonstrated until September 1873, when the *Grinder* again acted as the towing vessel and two of the runs were watched by members of the Admiralty's Torpedo Committee, which was reviewing the Harvey as part of its examination into the Whitehead, Lay, von Scheliha, and outrigger weapons. Unfortunately, during the final run, which was carried out at the highest speed attainable by the tug, one of the fuses exploded after twenty minutes and, on recovering

148 · *Nineteenth-Century Torpedoes and Their Inventors*

the torpedo, it was found that the insulated wire that formed part of the firing circuit had broken. It was not a serious malfunction, but the electrical system clearly needed to be strengthened. Any failure, no matter how trivial its cause, does not go down well when observed by VIPs.[52]

Very little more was heard about the towed torpedo in the press, and although used in service by the Royal Navy for a few years, it was abandoned in 1880.[53] Neither the United States nor Germany persevered with their versions for very long, and the French navy was the last to tire of the device, finally discarding it in 1886.[54] Some engravings showing the manufacture of the Harvey torpedo at Woolwich arsenal in 1877 have recently been discovered and are reproduced below.[55] So it seems that the government finally got its own way and withdrew the contract from the London Ordnance Works as it had first threatened to do in 1872.

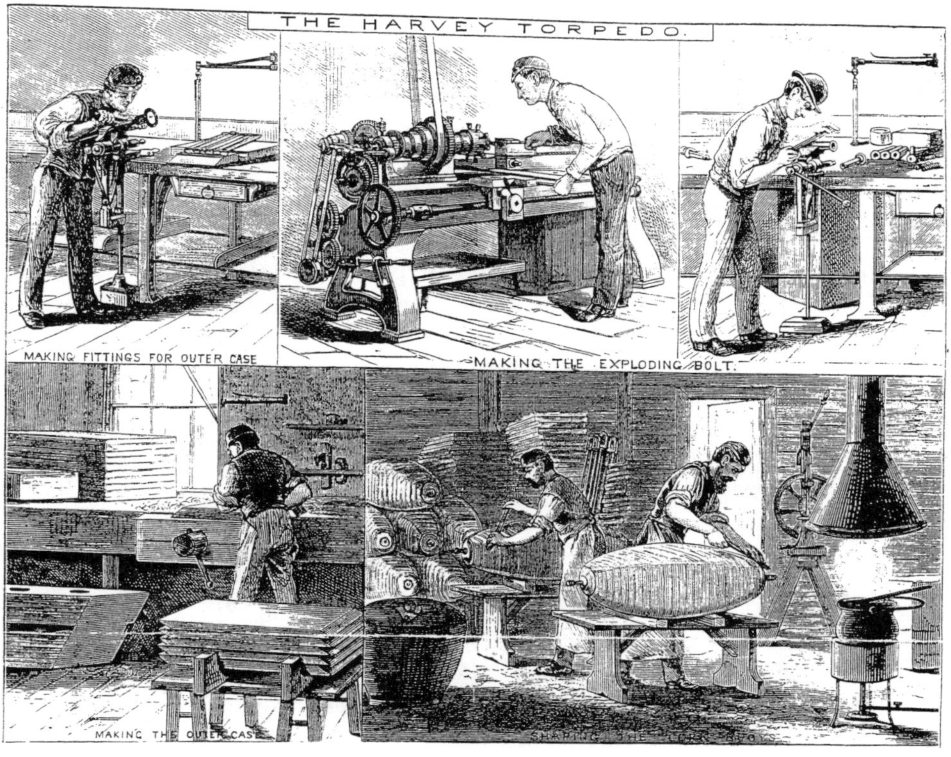

Manufacturing the Harvey towed torpedo at the Royal Laboratory, Woolwich.

A Perfect Nightmare · 149

As evidence of its unpopularity no patents have been traced for improvements to the towed torpedo. Captain Menzing's version of an electrically detonated copy of the Harvey has already received mention in a previous chapter and its abject failure in combat noted. Even Bushnell had seemed disinclined to do anything about the weapon after he had invented it. And yet, despite everything, it seemed a safer way to sink the enemy than the near-suicidal spar torpedo. If it had not had the misfortune of arriving on the scene contemporaneously with Whitehead's fish torpedo its future may not have been so bleak nor its life so short.

Reports on only two other towed weapons have been found. The first was a device "similar to the Harvey torpedo" which was tested on the Thames on Saturday, 4 August 1877. It had been invented by a mechanic from the Royal Laboratory by the name of Griffiths—his first name was not given. It was claimed to have superior steering control to the Harvey and could be guided to almost any angle to port or starboard of the towing vessel by the use of just two "light lines." No further details were provided, but the design was said to have been adopted by the Admiralty. Griffiths also demonstrated a telescopic spar torpedo the following Thursday. The precise function of this is not entirely clear from the text, and the relevant paragraph is therefore quoted in full:

> The spectators saw only a couple of poles, each about 30 feet in length, lying along the deck one upon the other, with a red disc at the extremity to represent a charge of guncotton, and the other end made fast a little astern of midships. The practice consisted of taking aim at the floating buoys in the river as the launch steamed past at full speed and, simple as the affair looked, the effect was remarkable. Steering within a calculated distance of about 50 feet, the torpedo was cast overboard, when the tide and motion of the vessel carried it out to arms length, and at the same time caused the upper spar to stretch out in telescopic fashion, carrying the torpedo's head completely under the object attacked.[56]

Judging by the description, the device would appear to have been more a "hinged" torpedo than telescopic. However, the invention was considered "both clever and valuable . . . and certainly to be preferred to the ordinary spar torpedo suspended over the bows of a vessel attacking end on, and risking its own destruction." Nothing further was heard of either invention.

The final example, which resembled in appearance an illustration from a Jules Verne story, appeared in 1889, nearly ten years after the demise of the Harvey, and was invented by a Frenchman, M. Lege. It was reported in the *Engineer*, though very few details were given. The towline was said to consist of one thousand yards of "fine wire." While this admittedly greatly increased the weapon's range, reliance on a thin-gauge thread of metal seemed to be unduly tempting fate. As the periodical commented, "We could only admit the feasibility of this particular [weapon] on its being shown to be so by a sufficient number of actual trials under service conditions."[57] The Lege also employed different tactics from the Harvey in that the torpedo was guided to the far side of the target and then hauled toward the towing vessel to strike the enemy's far side. The intention was to avoid any antitorpedo net defenses, but of course, if the target was protected by booms on either beam, the entire concept became pointless. The inventor claimed that he had "practically proved" he could observe the wire by which he was drawing the weapon, which seemed an overly optimistic statement with a submerged wire at a distance of a thousand yards and with the torpedo itself being on the blind side of the target.

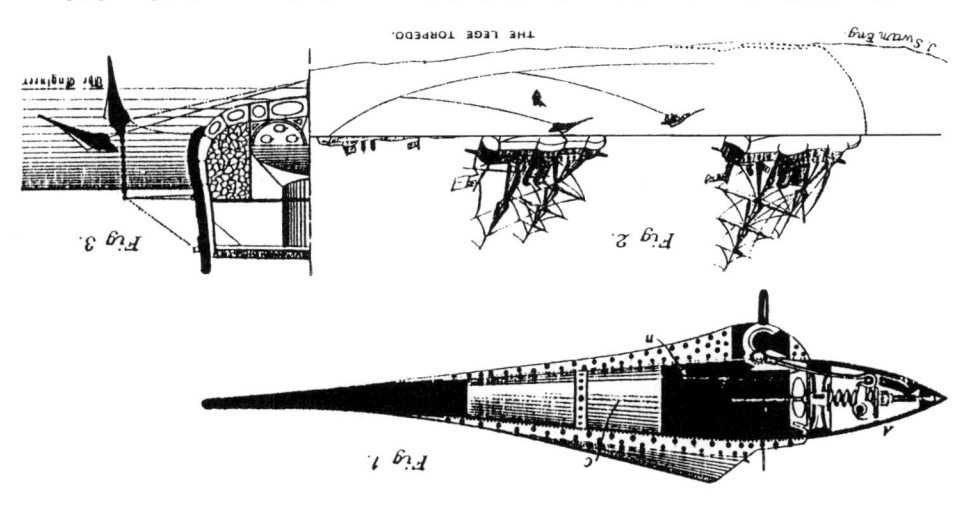

The fishlike Lege towed torpedo attacked the enemy's disengaged side, which meant the captain of the towing vessel could not see where it was.

The main part of the body was stated to be filled with cork, this "being a cheaper and more serviceable arrangement than any airtight construction." And although, according to the drawing accompanying the article, the contact trigger was in the nose, there seemed to be a total lack of stabilizing devices other than the upper and lower longitudinal fins. Neither is any apparatus provided to ensure that the weapon would always strike the target with its nose. In any event, thanks to the cork packing, it would have probably been bow-heavy.

Having examined the sorry saga of the towed torpedo, it is tempting to close the chapter on a similarly dismal note. But with the Harvey discarded and derided even by its erstwhile friends, Monsieur Lege's bold venture into these well-charted waters certainly confirmed the old proverb: Hope springs eternal in the human breast.

CHAPTER 8

The Wizard of Oz

With the sole exception of Louis Schwartzkopff, who built and sold large numbers of his pirated copies of the Whitehead weapon based upon plans acquired with the connivance of Adm. Alfred von Tirpitz,[1] the only man to make a substantial amount of money from the invention of a torpedo other than Robert Whitehead was an expatriate Irishman, Louis Brennan. Like his American counterpart, John Howell, his method of propelling the weapon owed absolutely nothing to conventional torpedo technology. Although born at Castlebar in County Mayo on 28 January 1852, Brennan was taken to Australia by his parents when the family emigrated to the Antipodes in 1861. He was barely nine years old on his arrival in Melbourne, and Australians, not surprisingly and justifiably, look upon his work–for he invented many other things in addition to his torpedo–with national pride and claim him as one of their own.

A great deal of extremely well researched material exists on the subject of the Brennan torpedo, the details of which will be found in the source notes for this chapter. While the number of Brennan torpedoes built can be numbered, at most, in terms of a few hundred, the production of the Fiume and Woolwich weapons ran into tens of thousands. The consecutive serial numbers for the Fiume-built torpedo alone had reached 13,167 by August 1914,[2] and to this total should be added the output of Whitehead's Weymouth factory, the British government's Royal Navy Torpedo Factory and its predecessors the Royal Laboratory and Royal Gun Factory at Woolwich, plus numerous official works throughout the world which built the Whitehead under license. Yet considerably more has been written about the Brennan than the vastly more prolific "fish" torpedo.

[152]

The reason for this is not immediately apparent, for Brennan's weapon was equally as much a closely guarded secret as the Whitehead torpedo. In many ways it had an even higher security rating than its Fiume rival, for, as a British government monopoly, it was never supplied to a foreign country, whereas the Whitehead company, with its aggressive marketing, sold to any nation with the money to buy. One aspect of the weapon which could account for this interest lies in the still extant ruined remains of the brick and stone forts that were constructed in the 1880s and 1890s as part of a coastal defense system based solely on the employment of Brennan's torpedoes for the protection of England's harbors and naval stations in a wide-ranging scheme that extended as far as Ireland, Malta, and even Hong Kong. These relics of past imperial glory have attracted the curiosity of industrial archaeologists, who have helped to swell the growing band of Brennan torpedo enthusiasts, and his antipodean background has aroused the interest of Australian historians justifiably eager to demonstrate their nation's contribution to nineteenth-century technological advancement.

With such a wealth of information available, this chapter must be restricted to an outline of Brennan's life, his inventions, and, of course, his torpedo. Although much new material is included, some of the facts given in the text have also, of necessity, appeared in *The Devil's Device*.[5] But as one of the very few successful torpedo inventors of the Victorian era, Brennan's weapon deserves close examination so that his achievement can be placed in its historical context.

Louis Brennan was educated in Melbourne, latterly at the Collingwood Artisan's School of Design, where, like Robert Whitehead at the Manchester Mechanics Institution some thirty years earlier, he studied engineering by way of evening classes. He was clearly an outstanding student for several of his inventions—a window-latch, a mincing machine, and a billiard marker are early examples of his work—were exhibited at the Juvenile Industries Exhibition, which was held in Victoria in 1873. Some time before this, he was apprenticed to Alexander Smith's foundry and workshop at Carlton. The latter was not only an expert engineer and a thrusting entrepreneur but also well placed in local politics as an elected member of the Victoria legislative assembly. Significantly, he was an officer in the Victoria Volunteer Artillery Regiment, which had close links to

its sister unit, the Victoria Torpedo Corps. These connections were to prove of considerable value to Brennan when his torpedo was shown to be a practical and viable weapon.

It has been claimed that the Brennan torpedo was the world's first guided missile, but as we have seen, Ericsson, Lay, and von Scheliha had all built and operated dirigible torpedoes controlled by shore-based operators using compressed air or electricity. The only difference was that Louis Brennan's weapon was much simpler in its basic concept and it ultimately worked perfectly over both an acceptably long range and at a satisfactory speed. In deference to his pioneer predecessors we should perhaps slightly amend the eulogy to "the world's first *practical* guided missile."

According to Robert Graham, a one-time colleague of Brennan at the Royal Aircraft Establishment at Farnborough when the latter was working on a helicopter of his own design and exploring rotor technology many years ahead of either Sikorsky or Focke-Achgelis, the idea of what was to be the torpedo's unique propulsion system came to him while he was daydreaming in front of a belt-driven planing machine.[4]

Brennan realized that only one side of the belt was transmitting work and that it would be possible for power to be transmitted through a single line for a limited period—the latter's duration being dependent upon the length of belting wound. In a nutshell, the harder the device pulled backward, the faster it would go forward.

The late Norman Tomlinson, one of Brennan's first biographers, related a slightly different account but one which is probably easier for the layman to understand.[5] Placing the incident in 1874, he claimed that Brennan had initially visualized the concept by noting that a cotton reel, if the thread is pulled toward the operator from underneath, moves forward, not backward. And it did not take long for the young inventor to realize that the only device requiring propulsion for a limited distance, and which did not have to make a return journey, was the torpedo.

On the basis of his observations, the seed of an idea began to germinate in his brain. He started making rough sketches of such a weapon so that he could build a scale model for practical experiments, and as the concept developed, he sought the mathematical assistance of W. C. Kernot (later Professor), a lecturer at Melbourne University.[6]

Digging into the depths of his very considerable theoretical knowledge,

Kernot produced a number of calculations of value to Brennan. One, for example, deduced that the speed of the wire through the water–or in other words, winding around the reel–would be double the velocity of the torpedo. A few years later, after the British government had agreed to purchase the torpedo, Brennan's company acknowledged Kernot's contribution by sending him a check for five hundred pounds.

According to Michael Kitson, Brennan's first patent was taken out in Australia in 1875.[7] This prototype torpedo had two propellers mounted side by side in a manner similar to that employed by John Howell's flywheel weapon. Thus, by varying the speed of the wires passing over each drum, the operator could cause either propeller to revolve more quickly or less quickly relative to the other, which would result in the weapon altering course to either port or starboard much as a ship's captain controls his vessel, without using the helm, by going ahead or astern on one of his engines (i.e., propellers). A primitive system of inlet valves and weights were incorporated for the purpose of depth-keeping. A wooden boat-shaped torpedo displayed at the Royal Engineers Museum at the Brompton Barracks in Chatham is thought to be this particular model which was tested a year later in 1876. However, as the internal mechanisms have been removed in their entirety, we cannot learn anything from it.

With Kernot's assistance, Brennan was able to improve both the depth-keeping and steering aspects of the wooden-hulled prototype, and by the beginning of the following year he had built a half-scale nine-foot working model which was demonstrated for the benefit of Col. Sir William F. D. Jervois and Col. Peter H. Scratchley, two Royal Engineers officers who had been sent to Australia to advise the various states' governments on coastal defenses. This demonstration torpedo weighed about 350 pounds and, at a speed of six knots, had a range of six hundred yards. Both men were impressed and recommended that the Victoria government conduct further trials. It seems possible that they reported their findings to London.

The specification of this improved torpedo had been lodged with the British Patents commissioners on 4 September 1877 and the requisite letters patent were sealed on 1 February 1878 with protection backdated to the date of application.[8] This new weapon varied considerably from the 1875 prototype. The hull was described as square in section and tapering in plan to a sharp bow and stern, according to one morbidly minded army officer "like a child's coffin."[9]

It was fitted with a single propeller, and steering control was now moved to a vertical rudder behind the screw. Although lateral steering was still described as being derived from speed variations between the two inboard drums, which reflected the rotating speed of the drums ashore, this was no longer achieved by twin propellers but by a complicated system of gearing linked to the rudder. Brennan had also modified his depth-keeping gear by the use of a hydrostatic valve which, while adequate at the slow velocities achieved in early trials, proved to be next to useless once speeds of fourteen knots and above were obtained by use of steam-powered winding engines.

Brennan continued to improve the weapon—a process of development that is probably responsible for the many errors contained in earlier accounts of the torpedo—and by August 1878, it had emerged with a length of fifteen feet with a fusiform (cigar) shaped body made from "best" iron boiler plate.[10] There were two other important alterations in design. The drums or winding reels were now placed one behind the other along the longitudinal axis of the torpedo and Brennan returned to his original scheme of twin propellers. This time, however, he adopted the familiar contra-rotating propeller system introduced to the Woolwich and Whitehead weapons by Robert Wilson in 1874.[11]

Early tests in the Williamstown Graving Dock were relatively successful. Steering proved to be reasonably under control, although the same could not be said about the torpedo's depth-keeping. Brennan, however, promised to improve this weakness.

The British Admiralty, aware of the army's interest in Brennan's weapon, had meanwhile instructed Rear Adm. J. Wilson, the commodore of the Royal Navy's Australian Squadron, to investigate the torpedo and report back. Alexander Smith, too, was working behind the scenes to obtain the Victoria government's backing for the project and raised the matter in the state's legislature on 2 October 1877. It is worth noting, at this stage, that throughout most of the nineteenth century, the British army—namely, the Royal Engineers—was responsible for England's shore defenses, the Royal Navy only being concerned with the island's seaward protection. Thus each service had different objectives and requirements when it came to weapon purchases, and at this point it was not clear into which category the Brennan dirigible fell.

Smith continued to lobby for financial support for Brennan's exper-

iments and was now able to add weight to his arguments by quoting the favorable impressions of the two British army officers who had examined it earlier. Nothing firm was offered, but it was apparent that the Victoria government was beginning to weaken in its resolve to do anything to help development of the weapon provided it did not physically involve giving anyone public money. But Brennan was now starting to obtain private finance for his work, notably from William Calvert. And it was due to these additional funds that he was able to build the 1878 improved model which would feature in the first public tests at Hobsons Bay, in the presence of the state's governor, on 21 March 1879. Senior army and navy officers, journalists, and members of the public were also invited–virtually the last time the Brennan torpedo was to be demonstrated in full view of the world. Within three years government security precautions had transformed the Brennan into a "secret weapon" to be concealed forever from prying eyes. This is no exaggeration. Even today, in the early twenty-first century, certain details of the Brennan's internal workings remain classified.

The torpedo put up a good, if brief, show, running for just over a minute at a speed of eleven knots before striking a small target boat at a range of four hundred yards. But like Whitehead's torpedo at the first Austrian trials of 1866, the Brennan demonstrated highly erratic depth-keeping and "porpoised" alarmingly during its short run. Fortunately for Louis Brennan, however, the military observers were adequately impressed by its performance and, as a result of a report submitted by the naval delegation present at the test, he was invited to come to England with his invention.

Important though the invitation was for Brennan's future, such trips cost money, and to help finance the journey the young Irishman formed the Brennan Torpedo Company. On its incorporation he and civil engineer John Temperley, to whom Brennan had assigned half of the rights in his patent in exchange for much-needed funds, were each issued with four hundred fully paid shares in return for surrendering their interest in the patent to the company. Further capital was provided by two other shareholders, Charles and Edwin Millar, and thanks to Alexander Smith's efforts Brennan received a grant of seven hundred pounds from the Victoria government shortly before the company's formation. Contrary to some accounts the Brennan Torpedo Company was incorporated

before Brennan and Temperley left for England and, as further confirmation, the registered office address in Melbourne was reported to the registrar general on 16 December 1880, just twelve days before they sailed from Australia.

Disappointment greeted Brennan and his codirector soon after their arrival in England following the Admiralty's examination of the torpedo, which was quickly found to be unsuitable for shipboard use. Bearing in mind that their lordships had invited the two men to Britain in the first place, it is difficult to understand how the Royal Navy's experts had failed to appreciate the basis of the torpedo's propulsion. The reports received from Australia must have indicated that the weapon relied on shore-based steam winches winding hundreds of feet of thin-gauge wire over drums, with the wire being drawn from similar but smaller reels inside the torpedo. To achieve a run of 350 yards, for example, 1,125 yards of wire had to wound around each drum. In fact, these vast lengths had been causing Brennan problems in Australia, for the local industry was unable to produce the necessary quality or lengths required. And just to make things more difficult, the wire imported from England had been cut into short lengths for shipping purposes. Brennan managed to join these together, but the maximum range he could achieve was only 350 yards. These joins naturally weakened the strength of the wire and, as it therefore had to be used with caution, only slow speeds could be obtained.[12]

It did not require a genius to appreciate that a torpedo which relied on wires and winches for its propulsion was scarcely a suitable weapon for a man-of-war. And it is apparent that the experts who considered the reports from their colleagues in Australia had failed to grasp this obvious feature.

Fortunately for Brennan and Temperley, the War Office proved more amenable than the Admiralty, which had, by now, lost interest in the device. And in early August 1881, a special Royal Engineer committee was instructed to inspect the torpedo at Chatham and report back directly to the secretary of state for war, the Rt. Hon Hugh C. E. Childers. This document was everything that Brennan could have possibly asked for and it strongly recommended that an improved model should be built at government expense. With typical bureaucratic caution the report was referred to a more powerful and influential committee, the members of which represented a cross-section of the army's most senior engineering

officers. This new body was equally impressed with the Australian weapon's potential and unhesitatingly supported the earlier recommendation. On 13 February 1883, the Brennan Torpedo Company signed an agreement with the British government under which five thousand pounds was to be paid immediately to cover all expenses thus far incurred by Brennan, while the inventor was personally engaged on a three-year contract at a mutually agreed fixed remuneration. As part of the deal, the government also provided a workshop at the School of Military Engineering at Chatham for Brennan's use, along with the necessary skilled labor.

It was at this point that a shroud of secrecy descended over the entire project; this has made it almost impossible for historians to determine precisely what improvements and modifications were introduced to transform the torpedo into a viable and reliable weapon. It was certainly a long and tedious process—often based on time-consuming methods of trial and error—and the depth-keeping problem was not solved until an indeterminate date, somewhere between 9 October 1885 and March 1886.

The men employed in the Chatham workshop carried out their duties in small isolated teams and were deliberately kept in ignorance of what the other groups were doing—a system favored by modern security services who wish to prevent individual agents from gaining access to too much sensitive information, though it is doubtful whether the workmen involved would have understood the mechanism even if they had seen it fully assembled and complete. To add to the air of mystery, the depth-keeping apparatus, when finally perfected, was supplied from the factory in sealed units or, as Millar observed, the Victorian equivalent of today's "black boxes."[13] And it is the contents of these latter items that have perplexed, and indeed frustrated, historians and researchers specializing in the development of the Brennan torpedo.

The lengths taken to protect the secret of the depth and steering control apparatus seem almost unbelievable, even in today's security conscious society. Nevertheless, Lieutenant Colonel Baker-Brown noted smugly in 1906 that "no foreign power—ever succeeded in copying it," thus implying that the precautions taken were both justifiable and successful.[14] On the other hand, of course, it is just as probable that no sensible foreign government thought the weapon worth stealing. What remains puzzling,

however, is why the last surviving Brennan torpedo, currently on public display at the Royal Engineer's Museum, still has both of its black boxes in situ yet, despite the passage of more than a century, the two sealed units remain unopened. Equally perplexing, the authorities have offered no explanation why they have failed to investigate and solve the riddle of their contents. According to one apocryphal story, an attempt was made some years ago to look inside the units by means of x-ray photographs taken through the steel outer shell of the torpedo's body. But to the dismay of the investigators, the exposed film, when developed, was completely blank.

The obsession with secrecy reached an almost farcical level when a standing order was promulgated requiring the sealed units to be removed from all torpedoes when the latter were not in use. Even worse, they were to be placed inside a locked safe, one of which was supplied to each launching station. It was, admittedly, a relatively simple matter both to remove and then to replace the black boxes in their respective weapons. But to implement this cumbersome procedure in the event of a surprise attack, such as the Japanese assault on Port Arthur in 1904, suggests a rather casual, almost unworldly, attitude to war by the senior echelons of

The British service version of Louis Brennan's unique wire-powered torpedo, which was probably superior to the Whitehead for coast defense.

the British army. Two different keys were needed to open each safe and both key holders had to be present to unlock the heavy steel doors so that the vital pieces of apparatus could be removed and fitted back inside the waiting torpedo. And until this drill was completed the weapon could not be launched against the enemy. The analogy of stable doors and bolting horses immediately springs to mind, although this was, more accurately, a matter of unbolting rather than bolting the proverbial portals.

The Brennan historian Michael Kitson noted that "in 1886 there were probably no more than eight people who knew how the depth and steering mechanisms functioned; four Royal Engineers officers who were directly responsible for its trials and development; two workmen who were building it at the School of Military Engineering; [plus] Brennan and Temperley."[15] Thanks to the diligence of Australian researcher Denis Cahill, however, it now seems possible that a ninth man was privy to the secret: Professor William C. Kernot. It will be recalled that Kernot had assisted Brennan in the very earliest days of the torpedo with the mathematical theory necessary to bring his unique system of wire propulsion to practical reality. And as already noted, the Melbourne professor later received an ex-gratia payment of five hundred pounds by Brennan's company—about twenty-five thousand pounds in today's terms—in recognition of his part in the weapon's ultimate success.

Millar confirms that Brennan had been in correspondence with Kernot in 1884 and 1885 concerning depth-control problems, "although the information he gave [to him] was very sketchy."[16] This may be so. But Cahill's discovery suggests that the mathematician possessed detailed knowledge of Brennan's solution to the difficulties of depth-keeping. And if his previous assistance to Brennan is anything to go by, he had in all probability made a substantial theoretical contribution to it.

The source of this important discovery was an obscure paper which Kernot had read to members of the Victorian Institute of Engineers at their Melbourne chambers at 178 Collins Street on Saturday, 7 July 1900. His subject was "A Proposed Improvement in Centrifugal Gravity Governors," a title apparently so irrelevant to torpedoes that it has passed unnoticed by the Brennan's many historians. Unfortunately, Kernot's actual text has not yet been traced, but an edited version was published in the institute's *Proceedings*.[17] Its importance is such that the relevant sections will be quoted in full:

A Proposed Improvement in Centrifugal Gravity Governors.

Paper read by Professor Kernot, July 7th, 1900.

While assisting Mr. Louis Brennan to work out his famous fish torpedo, which was afterward purchased for a large sum by the British Government, the writer had an instructive experience which has led him through after long delay to the subject of this paper. The torpedo was required to travel 10 feet below the surface of the water, and this it was attempted to secure by means of a horizontal rudder, actuated by a diaphragm exposed to the pressure of the water. When the torpedo was too near the surface, the diaphragm was pressed out against the water by a spring, and turned the rudder so as to steer downward. When too deep, the hydrostatic pressure overcame the spring, and the rudder was set to steer upward. At the proper depth the rudder was in its central or neutral position. The theory seemed all right, but the practical result was not successful. The torpedo had to be started at the surface, and at once disappeared, only to reappear at the surface about a hundred yards away, when it would dive a second time and reappear at 200 yards, and so on. In fact, it hunted abominably, travelling in a sinuous path between the surface of the water and a depth of 20 feet, or double that at which it was meant to go. The defect was ultimately cured by actuating the rudder partly by the diaphragm, and partly by a pendulum arrangement that took cognisance of the longitudinal slope of the vessel. Thus, if too high, but descending, the pendulum and hydrostatic appliances would counteract each other, and the rudder would remain at or near its central or neutral position. If too deep, but ascending, the same would be the case. If at the right depth, the hydrostatic appliance would be in its mean or neutral position, but the rudder would be entirely under the control of the pendulum, which would at once check any tendency to rise or fall. If too low, and descending still lower, or too high, and ascending still higher, both appliances would act in unison, and the rudder would act in its most powerful manner.

Now, the centrifugal gravity governor in any of its numerous forms takes cognisance of only one quantity, and that is the actual velocity of rotation, and its universally admitted defect is this, that it must either be made of such proportions as to be comparatively insensitive, or is made of proportions at all approaching those giving isochronism, hunts violently unless damped by the aid of a dashpot, or cylinder, with a leaky piston, and filled with oil or water, a confessedly clumsy and costly appliance. It re-

sembles the torpedo in its original and defective form. Now, I propose by an expedient of the most ridiculous simplicity, but which I have never yet seen suggested, to make it resemble the torpedo in its perfected form, and so abolish hunting, while retaining the maximum of sensitiveness and promptness of action.

To illustrate the method, let Fig. 1 represent a plan of a governor of this class; let A be the axis, which is, of course, vertical. Now, in the ordinary governor, the pendulum, or ball arm, moves in a radial plane, such as AB, and is affected by the absolute velocity of rotation only. Now, suppose a very long radial arm attached to the axis, and rotating with it, such as AC. On the end of this let us have a second pendulum capable of swinging in the plane CD, but incapable of any other motion. This second pendulum for small arcs will be sensibly independent of centrifugal force, and therefore of absolute velocity, but will be affected by acceleration or retardation; it will be like the second or pendulum appliance of the torpedo. If now the motion of the throttle valve or other means of varying the power of the engine be taken from both of these pendulums, we shall have a similar effect to that in the torpedo. If too fast, but reducing in speed, or too slow, but increasing in speed, the two pendulums will counteract each other, as they should on the principle of letting well alone. If the engine is going at the right speed, it will be under the control of the acceleration pendulum only, and any tendency to change speed will be nipped in the bud. If going too fast, and increasing in speed, or too slow, and diminishing, both pendulums would act in unison, and govern the engines with the greatest vigour. Thus we should have a governor taking cognisance of both speed and acceleration of both v and dv-dt, which is the real requirement. But a governor made as suggested, with two pairs of pendulums, would be clumsy and complex, though effective. It may, however, be made as simple as the present form by replacing the radial and tangential pendulums by one pendulum in an intermediate position, such as EF, which would be affected in the best proportion by both centrifugal and accelerating forces. What length, AE should be might be matter for experiment, and might be made adjustable in any given case, so as to be varied at pleasure. Unlike

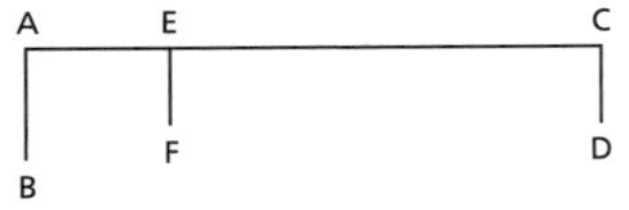

Fig. 1

the ordinary governor, this proposed one must be rotated in the direction of the arrow only. If reversed, the acceleration effect would take place the wrong way, and excessive hunting ensue. This, however, is no reason against its use, as engines with governors never do reverse.

It is apparent from this summary of Kernot's text that the key feature of Brennan's improved depth-keeping system was the pendulum. And an examination of Robert Whitehead's balance chamber, the Secret, reveals a marked similarity to its essentials. Whitehead's solution to the problem of depth control was described in *The Devil's Device* as "a movable disc fitted in the shell of the chamber and made watertight by a rubber joint. This hydrostatic disc, or valve, worked against a spiral spring and was either *in* or *out* according to the pressure of the water surrounding the weapon . . . with the aid of such a valve it was a relatively simple matter to preset the disc for a certain depth of water so that it sought equilibrium by operating a series of connected levers that raised or lowered elevator flaps on the [horizontal] rudders causing the torpedo to rise or dive until the point of equilibrium (i.e. the correct pre-set depth) was obtained."[18] Brennan's hydrostatic valve mechanism, as fitted to his earlier 1875 and 1877 models, appears to have fulfilled a similar function, although, like the pre-1868 Whitehead weapons, it suffered the same fault of "hunting" or "porpoising." In other words, both mechanisms failed to maintain a steady and constant depth.

The text continued: "A pendulum device incorporated into the mechanism compensated for any inclination or tilt of the torpedo and the movement of the pendulum weight acted as a further correction to the horizontal rudders and assisted the action of the hydrostatic valve by acting as a 'damper.'" It is immediately clear from Kernot's description that Brennan's pendulum system, which "took cognisance of the *longitudinal slope* of the [torpedo]," was identical in its working principle to the Whitehead balance chamber. The only significant difference lay in the fact that Robert Whitehead had invented the pendulum and hydrostatic valve method of depth control in 1868, almost twenty years before Brennan came up with the idea. As the Englishman had never patented his system, Brennan was, of course, perfectly free to "borrow" the concept. In addition, virtually every navy in the world possessed Whitehead weapons and, by that time, all torpedo officers were well acquainted with the mi-

nutest details of the Fiume depth-keeping mechanism. So what was the reason for this obsession with secrecy?

Pride is the first answer to spring to mind. It seemed likely that Brennan had pirated Whitehead's invention and wanted to conceal the fact, much as Robert Wilson never acknowledged his debt to John Ericsson's 1831 patent, which had originated the concept of twin contra-rotating propellers. But this was to do Brennan an injustice, for he was not responsible for the excessive secrecy that surrounded his weapon. The introduction of sealed units for the steering and depth-control devices was recommended by the Corps of Royal Engineers itself in 1886, very soon after Brennan had perfected his mechanisms: "The [Royal Engineers'] sub-committee felt there would be no difficulty in enclosing the more important parts in boxes which could be closed up at home [i.e., at the factory] and issued closed to the stations . . . [while they could be] returned unopened if they needed to be repaired."[19]

Kernot's statement that the depth-keeping defect of the earlier Brennan torpedoes "was ultimately cured by actuating the rudder partly by the diaphragm and partly by a pendulum arrangement" suggests that he was fully *au fait* with the details of Brennan's ultimate solution. In his thesis, Millar confirms that the two men were in correspondence at the time of the 1885 tests, and it is more than possible that Kernot either suggested or provided the mathematical basis for the pendulum system shortly before these took place thus enabling Brennan to perfect the idea a few months later. The wording of his paper certainly shows considerable familiarity with the initial problems arising from the torpedo's inadequate depth control and how they were resolved. However, the edited text of Kernot's paper contains several ambiguities which prevent a definitive answer to the riddle of the mechanism inside the black box.

The main point of difficulty arises when Kernot applies the principle of the torpedo's pendulum system to the subject of his paper, namely, the centrifugal gravity governor which he claims "[makes] it resemble the torpedo *in its perfected form.*" Then, having produced a diagram, he explained that "in the ordinary governor the pendulum, or ball arm, moves in a radial plane, such as AB, and is affected by the absolute velocity of rotation only. Now suppose a very long radial arm attached to the axis, and rotating with it, such as AC. On the end of this let us have a second pendulum capable of swinging in the plane CD, but incapable of any

166 · *Nineteenth-Century Torpedoes and Their Inventors*

other motion." Because it is concerned with the mechanics of the governor, I will not pursue his explanation further, except to highlight his reference to a second pendulum.[20]

Careful reading of the summarized text suggests that the edited version of Kernot's paper contains a confusion of expressions or terms which is unlikely to be found in the original and unabridged dissertation, for the professor was normally a model of clarity. In particular, use of the word "pendulum" could be misleading due to the difficulty of translating and reconciling the working parts of the torpedo's depth gear with the totally different requirements of a centrifugal gravity governor, even though the principle in each case is similar. In the process of condensation the editor, W. R. Rennick, the institute's honorary secretary, may have become confused by Kernot's mercurial reasoning. Thus the "first" pendulum of the governor's system would seem to equate with the hydrostatic valve mechanism of the original torpedo, hence the reference to "hunting."

This means that the "second" pendulum which was introduced to overcome the problem was, so far as the torpedo was concerned, the only pendulum involved and operated in conjunction with the hydrostatic valve as indicated in the earlier quote from Kernot's paper which referred to "a pendulum arrangement." To put the argument briefly, the torpedo had a hydrostatic valve and a single pendulum while the governor (which obviously could not employ a hydrostatic valve) used two pendulums, although Kernot seems to suggest that these could be combined together into one unit working in the plane EF as shown in his diagram.

And there, until and if Kernot's original text is discovered, the matter of the depth-control black box must rest. It seems safe to assert that it contains a pendulum arrangement. The ultimate conclusion, based on the facts presented in Kernot's paper, is in the hands of the reader. It needs only to be added that in 1897 the weapon was modified to operate with heavier gauge wire, which increased its range to 2,700 yards, considerably greater than either of its Fiume and Woolwich rivals.

Following the successful resolution of the depth-keeping problem, a further agreement was entered into by the British government on 18 January 1887 under which Brennan was to receive £110,000 (the equivalent today of at least £5 million) to be spread over a five-year period. He was also appointed superintendent of the new torpedo works at Gillingham in Kent at a salary of £1,500 per annum. Temperley, as his deputy, was to re-

ceive a yearly remuneration of £1,200. During the parliamentary debate that followed, it was further revealed that the capital sum was to be paid free of tax with an immediate payment of £30,000 to Brennan. The balance would be spread over the remainder of the contract's five-year term. Finally, in 1892, by way of icing on the cake, the Australian inventor was honored with the CB (Companion of the Most Honourable Order of the Bath) at the remarkably young age of forty.

Brennan's dirigible torpedo was regarded as the most important coastal defense weapon in Britain's military arsenal until 1906, when it was phased out and replaced by the new 9.2-inch gun. Between 1884 and 1894 a series of torpedo forts were erected, or built into existing fortifications, around England's coasts to store and operate the revolutionary wire-guided weapon, which, of course, needed rail systems for movement and launching, observation control points, offices and quarters, and engine houses for the big steam winches that reeled in the wire from the drums inside the torpedo itself. These torpedo stations are of interest in their own right, especially as the ruins of many can still be seen. They fall outside the parameters of this work, but the following brief details may help to serve as a reference point for readers who wish to explore further (the sources given in the footnotes may also be of assistance):

1. Garrison Point Fort. This station, built into existing fortifications, was situated at Sheerness and close to the Brennan Factory. Work on it began in 1884.

2. Cliffe Fort. Again built into existing fortifications, it was part of the Thames's defenses. Very few details of its construction history have so far been traced.

3. Fort Albert. Another station built into an existing fort. It stood on the north coast of the Isle of Wight facing Hurst Castle across the Solent. It is reputed to be the first completed Brennan torpedo station.

4. Fort Ricasoli. Situated in Malta and built into existing fortifications guarding the western side of Grand Harbor.

5. Fort Tigne. A purpose-built fort also situated in Malta and placed to protect the eastern entrance of Grand Harbor.

6. Cawsand Bay. Intended for the defense of Cawsand Bay in Plymouth Sound, it was built into the cliffs on its western side. Sometimes referred to as Pier Cellars.

7. Fort Camden. The only Brennan station to be situated in Ireland–at

that time politically part of Britain. Positioned to defend Cork Harbor, it was completed in 1893 at a cost of £9,225.

8. Hong Kong. Built at Lye Mun to guard the eastern entrance to Victoria Harbor and situated on the northern coast of Hong Kong overlooking the Lye Yue Mun channel.[21]

Further torpedo stations were planned, including one in Brennan's home state of Victoria, Australia, where it was to be situated at Observation Point. Only one of these projected stations, Dale Fort at Milford Haven, came into service, but it was employed for the use of the Zalinsky dynamite gun rather than the Brennan weapon.

Louis Brennan continued his association with the British government long after his torpedo was dead and buried, and during World War I and afterward, he was engaged in developing a helicopter many years before the better-known pioneers of rotor flight. Official funding ceased when the war ended, and Brennan finally abandoned the project, although it is claimed that the aircraft was actually flown by Robert Graham. Before this, he had constructed a gyro-stabilized monorail system, but although it worked efficiently and had lower operating costs than a conventional railway, it was rejected by both the British and Indian governments. He also demonstrated his considerable versatility as an inventor with the Brennanograph, a five-key typewriter similar in principle to the "speed-writing" machines used by stenographers in law courts and committee meetings where silent operation is vital. Louis Brennan died on 17 January 1932, just a few days short of his eightieth birthday, after being knocked down by a motorcar while out walking.[22]

John Adam Howell's early flywheel-propelled torpedo was examined in chapter 1, and this seems an appropriate point in the narrative to compare his career with that of Brennan before going on to consider the final development stages of a weapon that was, in its own way, just as revolutionary and unique in concept as that of his Australian counterpart and near-contemporary.

Born at Bath, Steuben County, in upstate New York on 16 March 1840, he entered the United States Naval Academy at Annapolis in 1854 and graduated with honors four years later. His early career reflected the typical routine of a young officer busily gaining practical experience of his

chosen profession aboard a variety of warships, including a spell of service abroad with the Mediterranean Squadron. Promoted to lieutenant on 19 January 1861, he was involved in several minor operations while serving with the West Gulf Blockading Squadron during the Civil War before returning to the Naval Academy in 1867 as head of the department of astronomy and navigation. And it was during this four-year period at Annapolis that Howell conceived the idea of the flywheel as a source of propulsive power, no doubt spurred on in his efforts by the reports filtering through from Europe about Robert Whitehead's new wonder weapon.

Indeed, in August 1869, having arrived at Fiume in the Mediterranean Squadron's flagship USS *Franklin*, Rear Admiral Radford had gone ashore "to study the effects of [Whitehead's] torpedoes" and was considerably impressed with what he saw.[23] But the asking price of twenty thousand pounds was unacceptable for a country still recovering from the financial aftermath of a civil war, and Radford withdrew empty-handed.[24]

Bearing in mind Howell's busy naval career, it is understandable why his flywheel torpedo took so long to develop and it is apparent that his duties must have forced him to neglect the weapon for lengthy periods. In fact, much of the progress toward perfecting the weapon took place when he was serving in shore-based appointments, notably his two tours of duty at the Annapolis Naval Academy (1867–71 and 1874–78) and his first stint as an inspector of ordnance for the Navy Bureau of Ordnance when he served at the Washington Navy Yard from 1881 to 1888, the latter date being of considerable significance, as will soon be explained.

It was during this first spell at Annapolis that Howell built his experimental model and was granted his 1871 patent. Thanks to this prolonged period ashore, he had sufficient time to modify his prototype and replace the original horizontal flywheel with a wheel that was perpendicular to the longitudinal axis of the torpedo. But further experiments had to be put aside when Howell was appointed for hydrographic survey duties, although his valuable contribution to geodetic science won him promotion to commander. The end of 1874 saw him back at Annapolis for another four-year spell, and this was immediately followed by two years at sea in command of USS *Adams*, which left him little time or opportunity to work on his torpedo. Conditions became more favorable in 1881, when he joined the Navy Bureau of Ordnance and served as an inspector at the

Washington Navy Yard. Promoted to captain in 1884, he remained with the bureau until 1888.

Confirmation that his naval duties had impinged on the time he had available for working on improvements to his torpedo is apparent from the date on which he obtained his patent for "an improved fly-wheel torpedo"–27 January 1885.[25] This meant he must have been working on the weapon in 1884 and, almost certainly, for several years before that. As the chronology of his career has shown, he had, throughout that time, been ashore and working for the Navy Bureau of Ordnance–a fortuitous posting for an inventor engaged on torpedo development.

Any further experimental work had to be shelved when he was given command of USS *Atalanta* in 1888, although he continued to find time to take out various other patents connected with naval ordnance up until 1892. It is possible that this spell of seagoing duties may have been a contributing factor in his decision to dispose of his torpedo patents to the Hotchkiss Ordnance Company of America on 1 February 1888. Certainly his responsibilities in the Navy were increasing year on year for, after a short period as president of the Steel Board, he was first placed in charge of the Washington Navy Yard and then the League Island, Philadelphia, establishment.

The Spanish-American War of 1898 found him at sea again in the rank of commodore as the flag officer of the Mediterranean Squadron, after which he led the blockading force that operated to the north of Cuba during the closing stages of the conflict. Howell had served his country well, and, promoted to rear admiral before the war ended, he finally retired from the active list in 1902. He died on 10 January 1918.[26]

The 1885 patent provides the only reliable guide to Howell's improved torpedo which was finally taken into service with the U.S. Navy in 1890–fifty having been ordered from the Hotchkiss company the previous year. It is interesting to note that, apart from the Brennan weapon examined in the first part of this chapter, Howell's flywheel torpedo was the only non-Whitehead design to go into series production. The Howell Mark I, as it was termed, thus joined the U.S. Navy four years ahead of Bliss & Company's licensed version of the Fiume 45-centimeter Mark I. These two facts alone are sufficient to confirm John Howell's important and influential role in the history of torpedo development.

But returning to the 1885 patent, it is clear from the specification that Howell recognized the gyroscopic effect of the flywheel and the important contribution it had on the maintenance of an accurate course. He had also now come down firmly in favor of positioning the flywheel with its axis horizontal to the longitudinal axis of the torpedo, an improvement he had only added as an afterthought in his 1871 patent. As these terms can be confusing without the aid of a diagram, and to avoid any misunderstanding, the relevant section of the patent reads as follows: "Inasmuch as a laterally-deflecting force tends to turn the torpedo about a vertical axis, it follows that the axis of rotation of the fly-wheel should be horizontal, and I generally prefer to have this horizontal axis of rotation transverse to the longitudinal axis of the torpedo." In general terms, his depth-control system now closely resembled that of the Whitehead Secret, which was, by that time, a secret no more.[27] "A diving rudder . . . is operated automatically by [a] mechanism whose action is controlled by a combined pendulum and regulator, the action of the regulator being governed by the pressure of the water in which the torpedo is immersed, which pressure of course varies with varying depth."

Apart from these two extracts, the six-page specification is too detailed to permit intelligible condensation, but it can be remarked that Howell at no point indicated the size of the flywheel or its speed of rotation, although a professional journal,[28] with reference to the 13.3-inch model, gives the weight as 110 pounds and the rotational speed as 10,000 rpm. While probably accurate, these data cannot be regarded as definitive. Two other features require mention. The 1885 model introduced the twin side-by-side propellers, although in his specification Howell observed that on occasions a single propeller would be adequate. The body shell, like most of his weapons, was made from brass.

The flywheel's initial rotation could be actuated by steam, compressed air, or electricity, although the most favored means appears to have been the Dow steam winch. It was claimed that it took only thirty seconds to fully charge the flywheel, but this was dependent on the power of the unit employed. If a small motor of only two horsepower was used, the time extended to a full five minutes. The 13.3-inch weapon referred to above was probably the prototype of the 1885 patent and in addition to the information already given the journal provided further, and more technical,

details. The radius of gyration of the 110-pound flywheel was given as 5.4-inches, which, at 10,000 rpm, produced an energy of 167 foot-tons. At maximum rotation, which was stated to be 14,500 rpm, this was increased to 347 foot-tons. Performance details (these are presumably not official U.S. Navy figures) of the 8-foot or 13.3-inch Howell were shown:

6400 rpm	18 knots for first 200 yards	Maximum range 500 yards
8400 rpm	24 knots for first 200 yards	Maximum range 800 yards

The propellers were also reported to have had a five and three-quarters inch diameter with a pitch of seven and a half inches and were "geared down 3 to 5 to the fly-wheels by bevels."[29] The report also included outline drawings of a 9-foot Howell with a diameter of fourteen inches and a one-hundred-pound warhead together with a 12-foot model seventeen inches in diameter carrying an explosive charge of two hundred pounds. No further information on the two latter weapons have been traced.

Another unofficial source, *The Naval Pocket Book*,[30] provides unconfirmed details of the "Model 1894" Howell, giving a diameter of 14.2

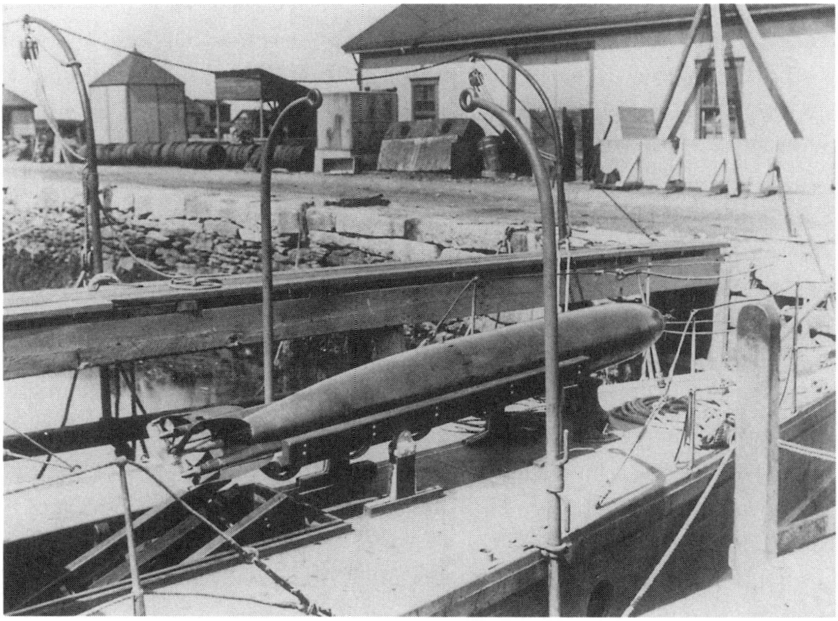

The Howell torpedo in a deck cradle, September 1892.

The Wizard of Oz · 173

inches and a length of eleven feet. Its weight was shown as 520 pounds with a warhead carrying 100 pounds of guncotton. Performance was stated to be twenty-six knots over four hundred yards with a maximum range of eight hundred yards. A "Model 1895" was also listed and shown as having a diameter of seventeen inches, a length of fourteen and a half feet, and a weight of 1,130 pounds. The warhead comprised 174 pounds of guncotton and the torpedo had a maximum range of one thousand yards. Performance was given as thirty-three knots for four hundred yards, thirty knots for six hundred yards, and 28.5 knots for eight hundred yards. As will be noticed these details differ slightly from those quoted in the *Engineer*, but it seems clear that there were fourteen- and seventeen-inch versions of the flywheel torpedo.

The same source gave details of official tests of the "newest model" which presumably relate to the Model 1895. The results make interesting reading. Out of 345 shots from a launching tube placed thirty inches above the surface, 332 hits were scored while 10 were regarded as misses and a further 4 were adjudged to be accidents, although it was not made clear what this meant. With a launching tube nine and a half feet above the waterline, another 69 shots were fired and 63 hits resulted—a success rate of more than 90 percent. Unfortunately, the key factor, the range, is not given, and that, of course, detracts from the value of the data, although the target was said to be sixty-three feet in length with a depth below the waterline of eight feet. As no equivalent test results are available for either the contemporary Fiume or Royal Gun Factory weapons, it is impossible to draw realistic comparisons.

Bearing in mind that the Royal Navy claimed to have reduced directional errors to only 3 percent, even before the introduction of the gyroscope there seemed to have been little to choose between the compressed air weapon and its flywheel-powered rival by way of accuracy. But the Howell could boast one unique advantage—the absence of an air bubble track on the surface to herald a warning of its lethal approach to the sharp-eyed lookout.

John Adams Howell was a career naval officer, and his official duties took up the major part of his working life. Inventing the flywheel torpedo was very much a part-time task—one might almost say a labor of love because it is clear that profit was not his aim. Whitehead, by contrast, was a professional engineer with no distractions outside his field of work. And

from 1868 he was concerned solely with torpedo development. It is also clear from his biography that, having produced the world's first self-propelled torpedo, he deliberately set out to make his fortune from it.[31]

Although the flywheel torpedo was finally ousted from the U.S. Navy by the Bliss-built Whitehead weapon—the last outfit of Howells being issued to the 11,346-ton battleship *Iowa* in 1903—even compared with the achievements of Ericsson, Fulton, von Scheliha, and Brennan, Howell deserves to be rated as second only to Robert Whitehead in the field of torpedo development. Had he been a dedicated full-time inventor like his English rival, who knows what he could have achieved?

CHAPTER 9

This Risk of Premature Explosion

AFTER THE excitement of flywheels spinning at ten thousand rpm, torpedo fortresses linked to electrical power grids, and weapons reliant upon steam winches and high-tensile wire for their propulsion, it is time to return once again to the more mundane yet equally strange devices which were thrust in front of the unsuspecting public by inventors eager for fame and fortune. One of the first examples came from Louis Brennan's adopted homeland, Australia, and made its appearance in 1885. Unfortunately, very few details of Captain McEvoy's weapon are known, but from the bare description available, it was not so much a torpedo as a damp squib.[1]

Driven by a clockwork motor which was wound-up externally, it was intended as a replacement for the spar torpedo and was launched over the bows with the aid of a firing tube and a spring. Described as having a length of eight feet and a diameter of ten inches, it boasted a range of two hundred yards with a fifty-pound charge. The proposed price of seventy-five pounds aptly summed up its value as a weapon of war, although the inventor optimistically added that this figure was dependent upon a minimum order of ten. For a full-blown torpedo it was remarkably cheap, but for a mere clockwork toy, it was considerably overpriced.

Next on the scene was Hudson Maxim, brother of Hiram Maxim, inventor of the machine gun. An explosives expert who subsequently worked as a propellants consultant for E. W. Bliss and Company, the firm that was to build the Whitehead torpedo under license some ten years later, Hudson has often been accused of pirating Brennan's design lock-stock-and-barrel in his 1885 patent. And, in fact, there seems to be some substance to that. Hudson Maxim used a similar wire and winch system

[175]

176 · *Nineteenth-Century Torpedoes and Their Inventors*

to that employed by Brennan but adopted a slightly different method of steering which involved the wires crossing over and around a pulley set into the tiller. It was inferior to Brennan's system of steering control because the rudder centered itself and, under certain conditions, the operator had to guide the weapon continuously.

Maxim also used water ballast as part of his torpedo's depth-keeping mechanism. Brennan had described a similar method in his specification, although, in fact, he never adopted it in practice. This has, rightly or wrongly, left the impression that Hudson Maxim had blindly included it in his plans on the basis of Brennan's narrative description. The accusations of plagiarism are, however, of little consequence. Maxim's dirigible torpedo was never built and nothing further was heard of it after the initial flurry of interest in 1885.[2]

That same year saw a revival of interest in the concept of the rocket weapon, when Washington Irving Chambers of Kingston, New York, was granted letters patent for his marine torpedo.[3] The advance in technology since the earlier rocket projectiles was very apparent in the patent drawings, and at first sight, Chambers appeared to have devised a workmanlike missile which was intended to be fired from a gun of similar caliber to that of the torpedo's diameter, though this latter measurement was not indicated. An examination of the specification of the patent and drawings suggests that, so far as the projectile element was concerned, the device might have possibly worked. But its weaknesses were exposed when it entered the water at the end of its flight, for the appended plans make it clear that Chambers envisaged the torpedo running awash during its final approach to the target. To overcome the disadvantages inherent in a surface weapon, he proposed a complicated, and probably unworkable, arrangement whereby the explosive charge, which was carried on the underside of the torpedo, would drop down so that it would detonate below the waterline of the enemy vessel: "The hooks are disengaged from the magazine and the latter is allowed to drop out of its chamber before exploding, so as to be exploded at a distance below the surface of the water . . . [this] distance is regulated by the length of the connecting-bars by which it is hung to the body of the projectile." There is no record of the device being built, and like Hudson Maxim's torpedo, nothing further was heard of it.

The American-designed Patrick dirigible weapon, which can be dated

This Risk of Premature Explosion · 177

to 1886 or, more probably, 1888, was one of the largest torpedoes ever constructed. It was not only built but physically tested in the water. The length of the weapon was no less than forty-two feet, although experts disagree on its diameter, which is shown variously as twenty-two and twenty-four inches, the larger measurement being that quoted in the Royal Navy's torpedo manual.[4] As the torpedo was tested by the British Navy, the latter dimension is probably the most reliable.

Like many other similar weapons of the late nineteenth century, the main body of the torpedo was suspended beneath a large float which served two purposes: it avoided the complications that ensued from running submerged and, as it was steered remotely by electricity via an insulated cable, its operator was able to keep it in sight throughout an attack so that its course could be altered as necessary to ensure hitting the target. As pointed out earlier in connection with other float torpedoes, it also gave the enemy ample warning of the weapon's approach, thus enabling an erstwhile victim to take avoiding action or, even better, sink it with a few well-aimed salvoes. The float itself was made from thin copper sheeting packed with waterproof cotton which, it was claimed, retained sufficient buoyancy to remain afloat no matter how often it was hit by enemy shells. Fueled by "compressed carbonic acid gas liquified by ice," which was fed into a six-cylinder engine, the ungainly weapon was credited with a speed of twenty-one knots over a range of two thousand yards.[5]

The whole torpedo weighed six thousand pounds, more than two and a half tons, and traveled submerged at a depth of some three feet beneath its supporting float. The warhead was said to contain two hundred pounds of explosives. The only piece of information of doubtful veracity related to the method of launching which alleged that the weapon was fired from a smoothbore underwater gun by means of compressed air. Bearing in mind the enormous weight and length of the torpedo, plus the fact that it was linked to a large copper float, and not to mention its electrical guidance cable, the reference is plainly ridiculous and the writer was obviously confusing the Patrick with Ericsson's projectile weapon or something similar. A moment's reflection should have made it obvious that to launch a torpedo which relied on a surface float from a submerged tube or gun was, in any event, not logical.

As we have already seen, the float-supported torpedo was certainly not

a new idea and it tended to be adopted by inventors without the skill to devise a workable means of depth-control but who, nevertheless, realized the importance of detonating the warhead below the target's waterline. One of the earliest examples so far traced was thought up by Henry Julius Smith of Boston, Massachusetts. It is of interest historically for two reasons: the use of electricity as the medium for controlling the torpedo via a wire from an external source of current, a very early instance of such a guidance system which Smith patented only five weeks after von Scheliha's English provisional protection was granted, and his even more important recognition of the warming effect of sea water on compressed air or gas. "As previously intimated, the water of the sea has free-play upon nearly the whole surface of [the compressed-air] reservoir," he explained in his patent specification. "This is for the purpose of supplying the heat abstracted from the compressed air or other gas. Moreover, if all sources of heat are cut off the gas would lose its expansive force and some gases would freeze. *The sea water takes the place of the fire under an ordinary steam boiler.*"[6] This was a very percipient observation for 1872.

Another, unbelievably grotesque float-supported torpedo emerged from Alfred P. S. Miller of New York City in 1887.[7] The preamble to his patent again included that telltale phrase "to enable others skilled in the art to make and use my invention" which usually indicated the presence of an untried drawing-board idea. And, as such, it does not merit any detailed comment or examination. In appearance it resembled a gondola or an engine nacelle from a Zeppelin airship. Suffice it to say that it was a monstrosity fully equal to James McLean's incredible contraption of 1879.

George H. Reynolds, another citizen of New York City, designed a "Submarine Torpedo Boat," wire-controlled by electricity, in 1881.[8] An awash weapon, it incorporated a special axe-shaped bow which was intended to prevent the nose from dipping too deeply when running at speed. It, too, employed circulating water to keep the valves above freezing point and this seemingly almost universal recognition of the heater principle appears strangely at odds with the failure of the Woolwich and Fiume engineers to discover the phenomena and exploit the benefits of warming compressed air until 1901. This facet of torpedo technology has been mentioned before. But the delay in applying the heater principle to the Whitehead weapon seems so incredible, I feel it does no harm to highlight it again.

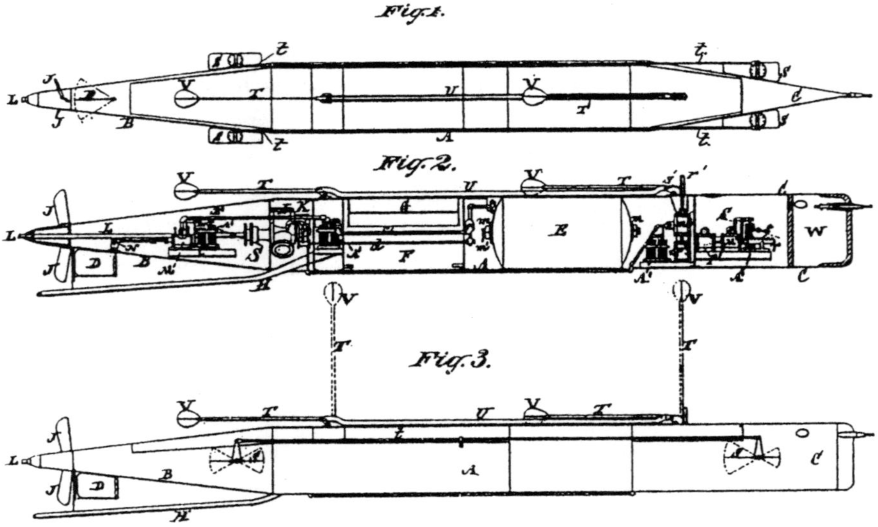

Fig.1.

Fig.2.

Fig.3.

George Reynolds patented this dirigible torpedo in 1881. Controlled by electricity via an insulated cable, it was fitted with electrically elevated "target rods" to permit visual steering from the shore.

Most of Reynolds's patent specification is concerned with the mechanism for raising and lowering the sighting rods, and the weapon is, in general terms, of little interest.

It is always rewarding when a new and previously untried idea is discovered, for many nineteenth-century designs bear a marked resemblance to one another. And the rotary, or rotating, torpedo patented by Sid Hugh Nealy, a resident of Washington, D.C., in 1887, certainly broke fresh ground.[9] His specification, unlike those of many other inventors, put the idea in a nutshell: "A motorshaft unprovided with propelling devices is held against rotation in the direction in which the motor tends to rotate it, thus producing a reaction which causes the rotation of the motor itself, which in turn drives or rotates the torpedo shell, having a suitable screw or propeller blade upon the outside." Readers familiar with early aircraft engines will immediately recognize a similarity with the principle employed in the French-built Gnome rotary motor, which first appeared in 1909, in which the cylinders and propeller rotated together on a common crankshaft. In Nealy's torpedo, however, the propeller was omitted and the weapon was propelled by the rotation of its own body shell.

To aid its progress through the water, the torpedo possessed a single fin in the form of a helix that ran the full length of the weapon's body in a continuous spiral, referred to by the inventor as a "feather." Unfortunately, he also described it as a "single screw propeller blade . . . extended from end to end of the torpedo," which at first reading seems to contradict his earlier statement: "I dispense entirely with propeller-blades." However, if one substitutes "Archimedean screw" for "single screw propeller blade," his intention becomes clearer. And despite the rather muddled wording, the sketch appended to the patent clearly illustrates the spiral fin or feather and confirms the absence of a conventional bladed propeller.

Sadly, Nealy's novel invention fell foul of two weaknesses which have already been noted in this chapter. He refers to "a motor, preferably a spring motor" which would have been similar to McEvoy's clockwork unit. And it is clear that such a method of propulsion would have lacked adequate power to drive the weapon for any worthwhile distance or at a satisfactory speed. He also employed that inevitable impediment so beloved by nineteenth-century inventors—a buoyancy float. This added a further complication for the torpedo had to be connected to it by means of rods and special swivelling linkage so that the weapon could rotate freely whilst running.

For some obscure reason, Nealy incorporated a sighting device, or hood, which rose above the surface at intervals so that "the observer may know the location and direction of the torpedo." This would appear to have been a pointless exercise on two counts. First, as the weapon, once launched, could not be externally controlled for either depth or direction, it mattered little whether or not the observer knew which way it was going, unless he needed time to dream up an excuse for missing the target before the torpedo actually did so. And, second, as the suspension float ran awash, it would have been visible to an attentive observer throughout most of its run. As a final refinement, a rod projecting from the front of the float released the torpedo as it came into contact with the target, thus allowing the explosive section to continue moving forward under its own inertia while, at the same time, in obedience to the laws of gravity, the spiral-finned torpedo moved downward through the water. Nealy claimed that this downward motion allowed the torpedo to pass below

the "guard-chains" of the antitorpedo netting. However, the patent's diagram, not to mention common sense, makes it clear there would have been insufficient time or space for this final maneuver. It is a pity that Nealy's novel design was not further developed, for these irritatingly minor faults could have been identified and eradicated.

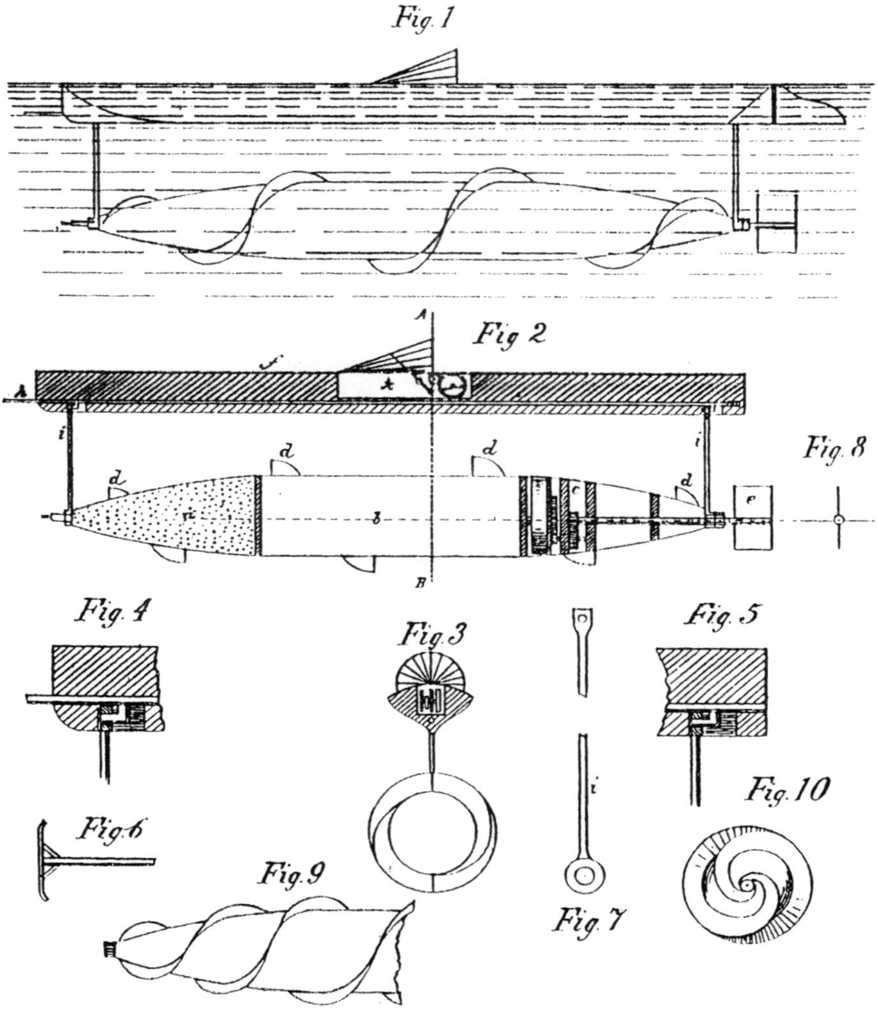

Nealy's unusual float-supported rotating torpedo, patented in 1887.

Nineteenth-century inventors seemed as fascinated by rockets as are today's English children on Guy Fawkes night and their American cousins on Independence Day. The same year that Nealy patented his clockwork rotary weapon, Timothy Sullivan and Ernest Etheridge, both from New York City, produced yet another rocket-powered torpedo.[10] Briefly, it was composed of two cylindrical compartments connected by a center section. The rear compartment contained the combustible rocket propellant, "which, in burning, rapidly generates a gas which is ejected rearward and acts upon the water and drives the torpedo forward with great force." It is a constant source of amazement how nineteenth-century inventors always made everything sound so simple, but perhaps there were advantages to not becoming too involved with detail. The forward section held the explosive charge inside, which was a small central section filled with dry guncotton powder. This primitive detonator was triggered by means of a cartridge which was thrust back when the spring-loaded firing pin struck the target. The "rocket composition"–the technology of rocket motors was still very much in the future–was ignited by a fuse or cap which was activated on launching. This latter operation was described vaguely as being "projected from a gun or similar device." Directional stability under water was supposedly controlled by four radially positioned rudders, the mechanics of which seemed rather obscure and were unlikely to have had any influence on the course of the weapon. It is doubtful whether this particular torpedo ever saw the light of day.

The year 1887 also produced what must be the most unusual torpedo ever invented, a weapon that carried no warhead and was not designed for military employment. Indeed, very few people could envisage a torpedo that was not intended to sink enemy ships. But Henry Morden Bennett did. And in so doing, he created a totally unique torpedo that, far from killing enemy sailors, was to be built for the express purpose of saving lives.

Despite taking out a U.S. patent, Bennett was an English clergyman from the seaside resort of Bournemouth in Hampshire, and the preamble of his specification explained that his invention was intended "for communicating with shipwrecked vessels . . . and of conveying a line, provisions, and life-saving apparatus to a ship in distress." The unarmed weapon also carried a means of "establishing telephonic communication" with the vessel's survivors, a remarkably farsighted concept mir-

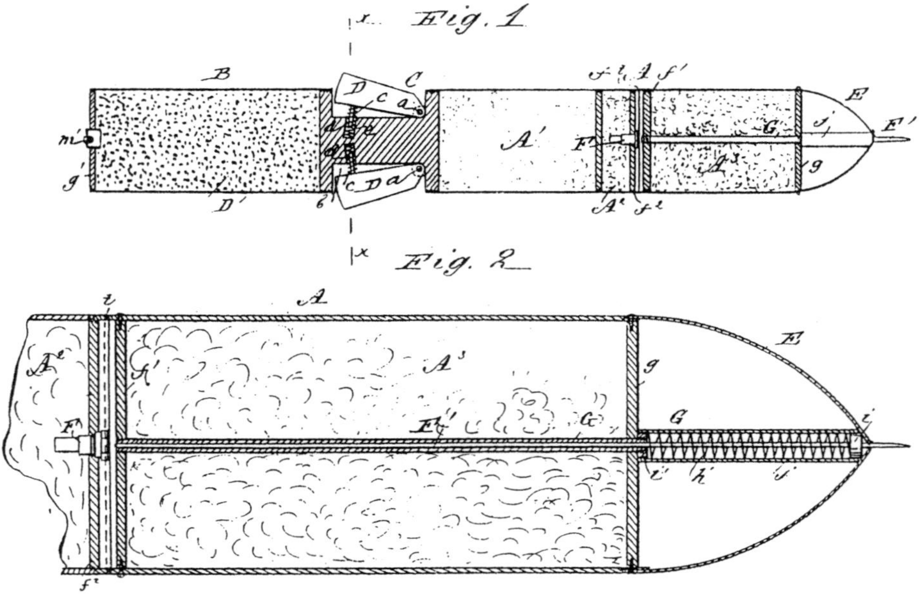

This Risk of Premature Explosion · 183

An interesting twin-bodied rocket torpedo invented by Sullivan and Etheridge in 1887.

rored today in unmanned diving bells and similar devices employed in submarine rescue operations.[11]

Bennett had obviously appreciated the difference between the free-running Whitehead-type fish torpedo and the electrically controlled dirigible weapon which John Lay and others had produced, and his apparatus was tailored to be used with either design. This, however, was not immediately apparent in his specification, for, having illustrated a Lay torpedo, his text described a "small" torpedo to be towed astern which would contain the control equipment necessary for steering the main torpedo via a cable link to an external operator, the latter being ashore or on board another vessel. As the Lay already had a similar built-in control system, the necessity of the towed capsule was puzzling. But on carefully rereading the specification it became clear that the towed capsule was only intended for use with the Whitehead. The Lay torpedo merely required modification so that it could carry and release the life-saving equipment which was to be carried in the warhead space, the inventor having recognized that the dirigible weapon did not require any additional controls.

184 · *Nineteenth-Century Torpedoes and Their Inventors*

In the patent Bennett described himself as "a Clerk in Holy Orders." It was appropriate, therefore, that he should invent a method of "turning swords into ploughshares" or, more accurately, "torpedoes into rescue vehicles." Although his invention was not taken up and developed, the germ of his idea remains alive today in the form of the life-saving rocket, although the latter had, in fact, started life in 1821.

Thorsten Nordenfelt, the Swedish armament's tycoon, has already been briefly mentioned when his name became linked with that of John Lay during the demonstration trials of the latter's dirigible torpedo for the Turkish navy in 1882.[12] Whether this encouraged him to proceed independently of Lay is not certain, but by 1888 Nordenfelt had blossomed forth with his own dirigible weapon which for sheer size nearly outstripped Patrick's monster. Indeed, its body diameter of twenty-nine inches was larger, although it came nowhere close to matching Col. Ottomor Gern's torpedo with its impressive girth of thirty-nine and a half inches. However, it weighed about one thousand pounds less than its American rival and was several tons lighter than its Russian counterpart of 1872. As its overall length is not recorded, it is not possible to make any other dimensional comparisons.

Accurate dating of Nordenfelt's dirigible is difficult, for he took out his first patent in England on 20 September 1883, and in the United States on 17 February 1885, although application had been made in April 1884.[13] However, the wording of both patents suggests that the torpedo had not yet been built at that time. Bethell contented himself by observing the date to be "some time before 1890" and the origin of the first-quoted date of 1888 has not been found. Sleeman, too, attributes its birth date to the British patent of 1883.[14]

Many writers refer to the Nordenfelt as a float-supported weapon, in one instance providing a sketch showing the torpedo suspended from two floats. Sleeman, however, disagreed: "The mode of submersion, peculiar to this torpedo, consists of two small wooden fins of special form, which are supported in the water by the hull; the latter possesses sufficient buoyancy to float itself in the event of the fins being knocked away." This observation was written in 1887, but the following year Nordenfelt obtained a further British patent, which I have not examined. However, there is no reason to suppose that the Swede had by then discarded the fins in favor of floats, and reference to the latter by journalists may have

been due to misinterpretation of the supporting drawings.

The wording of the 1885 U.S. patent is meticulous in detail: "The body has a center of gravity as far below the metacenter as possible and a very small reserve of buoyancy. If this proportion is, for instance, one three-hundredth, the body of the torpedo will then float with one three-hundredth of its volume above the surface of the water so long as the fin is not attached to it." The inventor then goes on to explain that "the fin, which is easily attached to the top of the body, is constructed of wood, cork, or other light materials, and can be covered with iron or steel plate . . . [and] the fin being placed on the top of the body its extra weight will cause the body to be entirely submerged . . . [so that it] will float with only a small portion of the fin above the surface." Thus Nordenfelt's torpedo was an awash weapon and did not rely on surface floats and negative buoyancy.

As a dirigible weapon, and no doubt influenced by John Lay, Nordenfelt used an external electric cable to stop and start the torpedo's motor. But the weapon's historical importance rests on its reliance on storage batteries for the power needed to operate its electric motors. Kirby states that an early version carried 108 storage cells which produced 18 SHP.[15] Sleeman, over eighty years earlier, had attributed the batteries to the Electrical Power Storage Company and added that the motor had been designed and built by Germany's Siemens factory.[16] Finally, Kirby gave the performance details as maximum speed sixteen knots with a maximum range of four thousand yards.[17]

The Nordenfelt dirigible weapon thus marks a milestone in torpedo development as the first to operate solely on battery-stored electrical power. The Williams torpedo of 1884 was a close competitor, proposing the storage of electricity inboard by means of accumulators, but Nordenfelt's British patent antedates this by a year.[18] And, of course, the Williams torpedo was never built and his concept of stored electrical power thus remained only theoretical.

Despite the Swedish inventor pointing the way ahead, fifty years were to pass before this new technology was developed and exploited. It was not until World War II that the electric torpedo, in the guise of the Kriegsmarine's G-7e, became a reliable combat weapon. Freed from the complexities of compressed-air power units, the simplicity of a battery-fed electric motor meant that a semiskilled labor force could be employed in

186 · *Nineteenth-Century Torpedoes and Their Inventors*

its production which, in turn, led to significantly lower building costs.[19] Today, at the beginning of the twenty-first century, there are now numerous high-tech power units employing both solid and liquid propellants. But the electric torpedo, of which the British Stingray antisubmarine weapon is a typical example, remains up there with the best. And dare we add, thanks to Thorsten Nordenfelt?

However, before becoming too euphoric, it is chastening to remember that when Nordenfelt built his first submarine with the Reverend George Garrett and sold it to Greece in 1886, he equipped it with a Whitehead torpedo. Admittedly his electric weapon was probably not fully completed by that date but Nordenfelt, as a level-headed realistic Scandinavian, no doubt realized that a weapon with a twenty-nine-inch diameter and weighing five thousand pounds was scarcely the ideal torpedo for use in a submarine. And adaptability was a key part of the equation when measuring a weapon's success.

The year 1889 witnessed the arrival of another electric torpedo, although this was, frankly, a retrograde design, as the current was supplied by cable from an external source. However, unlike either the Williams or the Nordenfelt, the Sims-Edison was actually embarked for operational use by the twenty-six-hundred-ton Brazilian auxiliary cruiser *Andrada* during the revolutionary war of 1893–94. But according to Bethell, it was never used "because there was no one on board who understood it."[20]

The technical press always referred to the weapon as the "Sims-Edison," a joint product of Winfield S. Sims of Newark, New Jersey, and Thomas Alva Edison (1847–1931), the latter a world-renowned figure for his invention of the incandescent lamp and the phonograph who, in his lifetime, was granted no fewer than 1033 patents. Sleeman, too, referred to it as "the Sims-Edison electrical torpedo."[21] But the two main patents of 1885 and 1891 were granted solely to Winfield Sims with no mention of Thomas Edison. No doubt there is a sound reason for this, but in this particular section on the trials and tribulations of the weapon, I employ the more familiar title of the Sims-Edison as used in the contemporary press, although perhaps Edison had no hand in it.

There was little difference between the 1885 patent and that of 1891, but it is of interest to note that an even earlier British patent was taken out in 1883 and that the application for the first U.S. patent was initially made on 29 May 1882.[22] It is, therefore, safe to conclude that the Sims

This Risk of Premature Explosion · 187

Edison went back to 1881 or before. Sleeman, however, states that "the first of these torpedoes was experimented with in 1877 and proved successful in the matter of controllability and submergence, but was deficient in speed, only some 6 to 7 knots being obtained with this weapon." He continued: "In 1880, with an improved weapon, a speed of 10 to 11 knots was realised, and this weapon has been subjected to a series of most exhaustive experiments under the direction of General Abbott, USA."[23]

By 1885, the torpedo had emerged as a dirigible (guided) weapon controlled electrically from a shore station. Like the Nordenfelt, the electric motor was built by the German Siemens Company and was said to draw 30 amps at 600 volts. Like so many of its close relatives, the Sims-Edison was supported in the water by a single float which was fitted with sighting rods. There were a number of detail improvements to the electrical circuits and relays which fall outside the scope of this brief survey but there were some fairly obvious modifications to the external appearance between the 1885 and 1891 models. The protective metal tube through which the electrical control cable passed was extended to abaft the propeller, a modification that suggested problems of snagging had occurred during trials, and the forward suspension stanchion was angled back 45 degrees to provide greater strength for the knife-edged forward brace while it was cutting through ropes or obstructions. The steering rudder

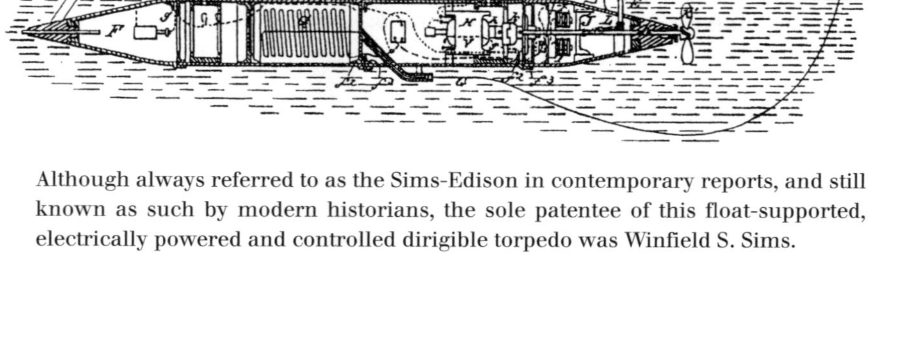

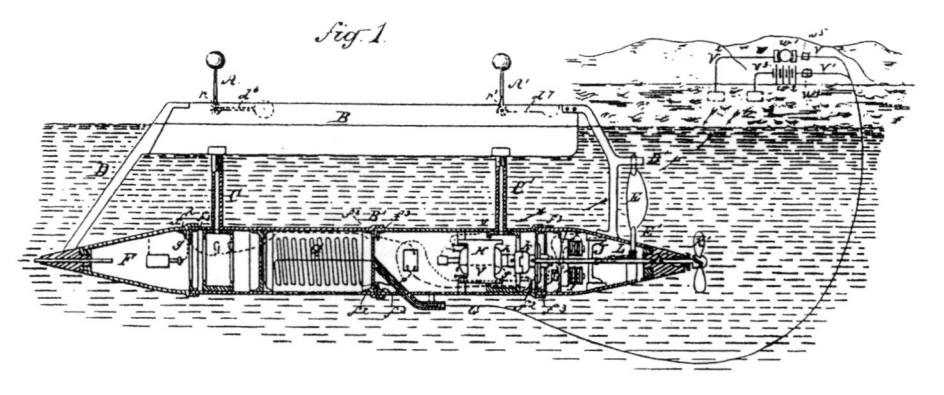

Although always referred to as the Sims-Edison in contemporary reports, and still known as such by modern historians, the sole patentee of this float-supported, electrically powered and controlled dirigible torpedo was Winfield S. Sims.

was also repositioned slightly and reduced in size. The speed was claimed to be twenty-one knots, although this seems optimistic in view of Sleeman's performance figures for the 1880 model, with a maximum range of thirty-five hundred yards, the latter figure being determined by the length of the umbilical electric cable, while the explosive charge was given as four hundred pounds. Unfortunately, it is not clear to which model, 1885 or 1891, this data refers.

The British trials, which were described at length in the technical press, served to confirm the problems created by this type of wire-controlled weapon, but they were particularly helpful in explaining how a float-suspended torpedo could be carried and launched at sea, and the following account is certainly not without interest.

The first tests of the improved 1891 model were held privately and, according to a report in the *Engineer* in January 1892, took place on the River Tyne above Elswick, home of the giant armament's conglomerate, Sir W. G. Armstrong, Whitworth and Company, which was clearly interested in the Sims-Edison, and in Stokes Bay, on the north side of the Solent near Portsmouth.[24] These were staged in advance of the official demonstration for which a small tug, *Drudge,* had been fitted out to launch the torpedo while the vessel was in motion.

The comments in the technical press reveal their continued opposition to the Whitehead torpedo, to which they now hastily added both the Howell and the Brennan. The Sims-Edison "is safer in the hands of its friends than the Whitehead or Howell, which may before launching be struck by a shot and exploded by means of [their] percussion firing gear. . . . The Whitehead has the further disadvantage of having half of its length filled with compressed air at 1000 to 1350 pressure to the square inch [*sic*] and a shot striking this could cause an explosion of air, if not worse."[25] This dire warning was then followed by a quite extraordinary statement: "This risk of premature explosion is so serious that high naval authorities have given as their opinion that in war no torpedoes would be fired from above water, but only from submerged tubes." As already noted, the first British demonstration of the Sims-Edison had taken place at Elswick, so Armstrong was presumably showing an interest in buying and developing it. By that time (1892), Josiah Vavasseur, who we last encountered as the owner of the London Ordnance Works, which had acquired the manufacturing rights of the Harvey towed torpedo, was one of the company's ex-

This Risk of Premature Explosion · 189

ecutive directors. His close connection with Fisher has already been re-marked upon. Could the "high naval authority" quoted by the *Engineer* have been the ubiquitous Jacky Fisher, who later that year was to become a lord commissioner of the Admiralty and comptroller of the navy (having already achieved flag rank as a rear admiral in 1890), pulling a few strings in support of his friend? Fisher proved adept at manipulating the press in his later career by means of what are, today, termed as "leaks," and it is more than a little odd that some of the more unusual causes he espoused seem to have had a vague connection with Vavasseur. It is interesting to note that *all* the dreadnought- and superdreadnought-type battleships that he created had submerged torpedo tubes. Yet, at the same time, Jacky led the race to build bigger and better destroyers, another of his conceptions, which carried their torpedoes in tubes mounted on open decks. But with typical ruthlessness, Fisher regarded them as "expendable," and the dangers, real or imagined, were ignored.

The same report was equally damning in its comments on Brennan's torpedo, and there is no question that the press, in its self-appointed role as guardian of the public purse, never forgot nor forgave the Australian inventor for the enormous amount of money he had been paid by the British government for his winch-and-wire-powered coast defense weapon. And it seized on every opportunity to criticize and denigrate his torpedo.

The much heralded official demonstration of the Sims-Edison took place at Spithead on Wednesday, 3 February 1892, in front of HRH the duke of Connaught, one of Queen Victoria's younger sons, who was, incidentally, a senior army staff officer, although he had admittedly trained at the Royal Military Academy at Woolwich and, as a subaltern, had served in the Corps of Royal Engineers. It was also significant that all the distinguished officers who accompanied the royal party, were military men, whereas senior members of the Royal Navy were conspicuous by their absence. Clearly, float-supported torpedoes were not considered seamanlike by Britain's admirals.

Although the demonstration was timed to begin at 2:30 P.M., *Drudge* had to leave at 11:30 A.M. so it could anchor off the Spit Fort and prepare itself for the exercise. "Here the final touch was put on the preparations by connecting the extension beam of the traveller rail, which projects some distance from the ship's side, and cannot therefore be conveniently

rigged out while in a crowded harbour like Portsmouth, for fear of fouling other vessels," the *Engineer* reported.[26] Although not described as such, this inconvenience was clearly a drawback to the weapon if it was to be operated at sea with other ships in close company or in bad visibility.

By 2:30 P.M. the aptly named *Drudge* was off Fort Monckton, where it embarked the duke and his party "with the punctuality for which the Royal Family are noted," and His Royal Highness was introduced to Winfield Sims, who explained how the torpedo worked and demonstrated the complexities of the control switchboard. He was also shown the cable which was paid out from the vessel's stern quarter through a projecting pipe which extended far enough to keep it clear of the vessel's propellers. Then, having put the duke fully in the picture, Sims gave the tug's captain permission to proceed and the little steamer headed toward the Spit Fort with black smoke streaming from its funnel. On receiving the order to "launch," one of the crew cut the spunyarn stop and the carriage began to run along the overhead beam carrying with it the torpedo, the propeller of which was already rotating. When it reached the end of the beam, an automatic tripper was actuated and "the torpedo dropped into the water with a slight dive, going well clear of the ship's side . . . [and as the current was increased by the operator] . . . at once gathered way and started off on its journey." Sims then demonstrated his ability to steer the weapon with a dazzling display of rapid course changes and even a serpentine S curve.

Up to this point it would be easy to forget that the report is describing the antics of a float-supported device, but having observed that the sea was "very rough," the writer continued: "There was no difficulty experienced in launching and controlling the torpedo which cut its way right through the opposing waves with a force which at times almost completely buried the float." Indeed, the only adverse comment was of an indirect nature and probably not even intended. For having described the departure of the duke and his staff, the report continued that he had "left the crew to pick up a mile and a half of cable, which was completed by 6 P.M." One can picture only too well a group of grumbling and weary sailors manhandling more than a mile of cable out of the freezing February sea with their numbed and red-raw hands as the royal party left for the naval dockyard and the convivial warmth of the officer's mess before returning to London in their comfortable steam-heated private railway carriage.

This Risk of Premature Explosion · 191

A further demonstration was organized in Stokes Bay on Tuesday, 9 February, and the *Engineer's* report this time revealed that, in addition to the English tests, Sims had taken the weapon to France a year or more earlier, where it had achieved twenty-one knots while being shown to the French authorities at Le Havre. In his account of torpedo development, Kirby suggested that various versions of the torpedo were produced.[27] While this may well be correct, the contemporary report in the *Engineer* gives the impression that there was only one demonstration weapon which was being constantly modified and tinkered with as Sims tried out new ideas: "It should be noted . . . that this torpedo has been used continually, and much knocked about during the last four years, besides which several additions have been made to it, to enable it to be conveniently lifted and launched from a ship." The torpedo still failed to achieve more than eighteen knots, but this particular demonstration was reported as "very successful in every way."[28]

The next trial of the overworked weapon took place the following Friday "[and] is to tell of failure," the *Engineer* reported coyly. It was again held in the far from sheltered waters of Stokes Bay and was attended by a large assembly of military officers headed by Major General Grant. On this occasion, three members of the staff from the Royal Navy's torpedo establishment, HMS *Vernon*, also deigned to be present. Louis Brennan, too, was another unexpected spectator who, we were informed, was greeted cordially by Winfield Sims.[29]

Unfortunately, and with the added embarrassment of Brennan's presence, the demonstration was not an unqualified success. The torpedo failed to move forward after being launched into the water and a short-circuit was diagnosed. Sims calculated that the fault was close to the junction between the "torpedo cable and the ship's cable" and he ordered the former to be cut near this junction so that he could connect the end to the control switchboard and thus continue with the demonstration. But the problem turned out to be less simple than was at first surmised. The tube through which the inboard end of the cable passed to carry it clear of the ship's propellers, which has been described previously, was apparently removed after the Elswick trial as it was deemed to be unnecessary, and the vulnerable cable's sole protection now rested on a metal guard around the torpedo's screw. This safeguard had, however, proved to be inadequate because a larger diameter screw had been fitted before this

particular demonstration and the new propeller had cut the all-important control cable.

Having lifted the dirigible out of the water and sorted out the immediate problem, Sims found himself facing a fresh setback a short while later when, during the next run, the steering failed completely and the weapon swung away in a long uncontrolled curve before being slowed and stopped by the operator. This time it proved to be no more than a loose connection on a secondary battery fuse, but it was sufficient to bring the demonstration to a premature end. The VIPs reembarked on the steamer that had brought them out from Portsmouth and, in the words of the *Engineer*'s reporter, they "returned to harbour; no doubt fully impressed with the idea that the thing is no good."[30]

The final demonstration of the series took place on Monday, 15 February, in front of delegations of naval and military attaches from the French, German, Russian, Italian, Turkish, Austrian, Spanish, Chilean, and U.S. embassies who had been brought down from London by a special train at the expense of the European Sims-Edison Electrical Torpedo Company. The inclusion of Edison's name in the company title confirms his connection with the business side of the torpedo and helps to account for the press referring to the weapon as the Sims-Edison. Nevertheless the fact remains that only Winfield Sims is named as the inventor in the letters patent. Stokes Bay was as the venue again and the little *Drudge* once more acted as the launching vessel. Much to the relief of all, especially Sims, the torpedo performed faultlessly, although, for the second time in four days, the final run ended prematurely when the maximum length of cable was expended. This, however, was a handling problem and no technical malfunction was involved. The weapon's speed was reported as eighteen knots against a two-knot tide, bringing it close to the twenty-one knots achieved in France, and the current employed was said to have been 25 amps at 950 volts, which produced approximately 24 SHP (shaft horsepower).

In its subsequent editorial, the *Engineer* performed a curious volte-face.[31] For having previously lauded the merits of the Sims-Edison, it now became highly critical of the entire concept and compared it unfavorably with both the Whitehead and the Brennan. Having pointed out the inherent disadvantages of using wires to control the weapon, and in some cases even to provide its propulsive power, with the ensuing risk of them fouling

This Risk of Premature Explosion · 193

the propellers or otherwise breaking under the strain of the forces to which they were subjected, the editorial continued: "A controlled torpedo is inadmissible as regards [use by] battleships and cruisers . . . and trailing a wire behind is sufficient to cause its rejection by the Navy as a ship weapon for general use." It also focused on the problem of observing a guided torpedo in combat: "[Although] it is often stated a torpedo can be controlled and directed beyond the limits of [defensive] machinegun fire, practical experience has not hitherto demonstrated the fact. Its effective action may be considered at the outside as 1500 yards. The difficulty is to know how near a torpedo is to the object. . . . A slight miscalculation would render the attack abortive. As the torpedo runs below the surface the projection above to show its position . . . is not a conspicuous object to those guiding the torpedo, and is only visible in clear weather. . . . At night the difficulty of directing the torpedo successfully would be much increased." Then, having considered various other disadvantages, the editorial warned of "the blunder of acquiring some new weapon of such limited value as the controlled torpedo" and concluded in forthright terms, "Their want of mobility, limitation of range, and difficulty in effective use, render them in all cases inferior to the locomotive torpedo carried in fast boats, and manipulated with skill and enterprise, examples of which have been frequently given in our naval manoeuvers."[32]

For some reason, the Sims-Edison had finally brought home to the armchair pundits the folly of their continual support for the controlled torpedo, the alleged advantages of which they had been espousing ever since the advent of the Whitehead weapon in the late 1860s. In fact, it had been more a matter of not seeing the wood for the trees than poor judgment. For it was not the concept of the controlled torpedo that was wrong; it was simply, as was so often the case in the nineteenth century, a lack of suitable and effective technology to make it work. Today most torpedoes have target-tracking systems, sometimes passive and sometimes active, and a large number are equipped for wire guidance. To the Victorians such marvels would have been found only in the pages of science fiction, a genre that, gaining credibility with writers like Jules Verne, was beginning to grip the public imagination.

Truth to tell, the inventors of the nineteenth century had the right ideas. But they would insist on running before they had learned to walk.

CHAPTER 10

An Eminently
Humane Weapon

WILLIAM K. CAVETT of Pittsburgh, Pennsylvania, had clearly not moved with the times when he patented his "Submarine ram and torpedo exploder" in 1890.[1] The fact that the self-propelled torpedo had, itself, put an end to close combat at sea, while the introduction of the quick-firing gun had transformed ram and spar attacks into suicide missions, did not deter Cavett from proposing "a ram which is adapted to be impelled rapidly and with great force against the bottom and sides of an opposing vessel below the waterline, and to convert the same into a torpedo exploder." In other words, a combined ram and spar torpedo in which the latter came into play after the ram element had done its dirty work. It scarcely needs adding that the familiar caveat "Such as will enable others skilled in the art . . . to make and use the same" was included in the patent's text. This usually meant that the inventor had not only failed to build the device but preferred that someone other than himself should place his life at risk by testing it.

It was certainly not made clear where the explosive charge was to be placed, and the warhead was not included in the patent drawing. According to the text, there appeared to be two options: the spar should be projected through a second orifice alongside the ram-shaft in the bows or it should be carried on a smaller shaft inside the shaft operating the ram. It is virtually impossible to sort out how this double-edged weapon was intended to work in practice as almost every alternative appears to be flawed. The answer to how the device actually worked is simple: it didn't.

George Read Murphy's "Victoria" torpedo, patented the same year, was an altogether more intelligent idea in which the Melbourne-based Australian inventor attempted to convert a Whitehead-type torpedo into

[194]

an electrically controlled dirigible weapon.[2] It received considerable coverage in the English technical press and was to be built in two versions— a shore-launched torpedo for coast defense and a "light naval" weapon for use by ships on the high seas. The former was described as being twenty-four feet in length with a diameter of twenty-one inches.[3] In view of Murphy's constant references to the Whitehead, it is interesting to note that the Royal Gun Factory, the successor to the Royal Laboratory, did not produce a 21-inch weapon until 1908. This had a body length of seventeen feet and ten and a half inches, although a "long" version measuring twenty-three feet and one and a quarter inches was produced the following year, surprisingly close to the dimensions of Murphy's prototype. The Fiume factory did not build its first 21-inch model (serial no. 9131) until September 1909, and the consignment ledger records it as being 6.3 meters, or 20 feet 7 inches, in length.

It must be concluded from these details that Murphy intended to construct his own weapon, and, indeed, the *Engineer* reported that the torpedo was in the process of being built by Heenan and Froude, of Aston, near Birmingham.[4] Official records, however, do not support the press comments. A board, headed by Major Ellery, was appointed in Victoria, Australia, and given the task of evaluating the Murphy torpedo, and the first paragraph of its report, dated 30 October 1891 noted, "It appears that this torpedo has never passed the theoretical stage, although it had for a considerable time been before the Imperial War Office and the British public." Having pointed out that the board had carefully examined everything put before it by the inventor, the report continued: "Although ingenuity and skill have been displayed, the Board does not see any marked superiority over what has already been developed in other torpedoes."

Dealing with Murphy's intention to anchor the torpedo and release it when the operational situation demanded, the board dismissed the idea out of hand: "This claim is entirely based on theory. From knowledge of the practical difficulty of maintaining air at high pressure, and . . . of maintaining electrical circuits when submerged for any length of time, the Board considers that this claim should be based on practical trial before being entertained."

The report also indicated that Murphy wanted to fit his control mechanism into a Whitehead torpedo "at the expense of the government," a request that seems to confirm that the 21-inch weapon allegedly being

built by Heenan and Froude had never got off the ground, if such a metaphor can be used in connection with a torpedo. The comments that followed this disclosure were no less scathing: "Mr Murphy proposes to obtain high speed by using very high air pressure, i.e. 2,000-pounds per square inch. It appears that a suitable air chamber for a locomotive torpedo, capable of standing this pressure, has not yet been constructed." Preempting the verdict of the press on the Sims-Edison weapon, the board observed, "At present it seems generally to be recognised . . . that a controlled torpedo is undesirable on board ship." The report concluded, "The Board cannot recommend to the Defence Department to incur the expense of a trial of this torpedo in its present stage."[5]

Despite Murphy's press statements that he was building an electrically controlled Whitehead torpedo—although a Freudian slip led him to use the name Whitbread (a famous English brewing company) in his U.S. patent[6]—both the drawings reproduced in the *Engineer* and his U.S. patent revealed the weapon to have had a float secured to the upper dorsal section of the body, a seemingly unnecessary addition in view of the Whitehead's balance chamber.[7] In the patent, Murphy states, "I dispense with the depth registering [*sic*] apparatus and . . . the horizontal rudder."[8] But he gives no reason for doing so. It is impossible not to conclude that Murphy found his own torpedo did not work—hence his request to the Victoria government for the use of an official Whitehead weapon for further experiments with his electrical control system. In view of his intention to disembowel it as indicated in the patent, it is not surprising they turned down his request. In any event the board's outspoken report had spelled the end of the Victoria Torpedo project and nothing further was heard of George Read Murphy.

In September 1892, the *Engineer* published a drawing of Jaque's detachable ram and subaquatic projectile "designed for the Ericsson Coast Defence Company" along with an article based on a paper he had submitted to the Iron and Steel Institute.[9] Despite the journal's attribution of the invention to Jaque, an absolutely identical weapon was patented by the Ericsson company in England in September 1890[10] while another exact duplicate of the design had been patented in the United States ten weeks earlier[11] by Valdemar Lässoe, who showed that the Ericsson Coast Defence Company had been assigned a one-half interest in the invention. Both patents were mainly concerned with improvements to the auto-

matic steering of the weapon after it entered the water, and, so far as can be seen, both texts were identically worded.

Jaque's paper, though submitted two years later in 1892, also refers to "automatic steering of the projectile in a vertical plane, after it enters the water." The relevant wording in both of the 1890 patents reads, "An automatic steering of the projectile in a vertical plane after it enters the water as to insure a true horizontal trajectory at a predetermined and fixed depth."

It seems likely that the report in the *Engineer* had confused the facts of Jaque's paper and had wrongly given him credit for the invention when he was, possibly, speaking on behalf of the Ericsson Coast Defence Company and describing the 1890 patents.

However, building bricks from straw must be guarded against, and there are too many assumptions involved to arrive at an explanation of this extraordinary muddle. It would be unwise to draw even a hypothetical conclusion from the inadequate evidence available. The only indisputable fact to emerge from this strange episode is that the Ericsson Coast Defence Company remained in business after the death of its founder.

Rather appropriately, a scale model of Alfred Johnson's dirigible torpedo was shown to the public at the Hamilton Juvenile Exhibition on 17 December 1891–to be absolutely precise, on "Thursday afternoon at 3 o'clock." It was described in the exhibition's bill matter as "A Masterpiece of Ingenuity."[12] Johnson, a clockmaker by trade, lived in Hamilton, Victoria, and was noted in a local newspaper report as "an inventor of no ordinary ability . . . [who has] recently attracted public attention with his 'Bushman's Watch.'" The torpedo model was said to be about four feet long–in one-third scale this would have made the prototype around twelve feet in length–cigar-shaped in form, with a rudder and two-bladed

ASROC technology of 1890. An artist's impression of Ericsson's projectile torpedo as improved by Valdemar Lässoe in his patent.

propeller at the stern. These last two features were controlled electrically by means of a cable connected to an external keyboard powered by a battery, although this would be replaced by a dynamo in the full-scale version. The stored electrical power also provided the propulsive force which, the report added, drove the weapon "through the water at a terrific rate." Not to be outdone in superlatives another newspaper drew attention to the "immense speed by which it can be propelled with the use of electricity."[13] To complete the weapon's specifications, the reports indicated that the weapon's impact detonator employed nitroglycerine while the main explosive charge consisted of guncotton "saturated with nitric acid."

Johnson told a journalist that he had submitted plans of his torpedo to the Victoria Defence Department, which had "desired him to construct a working model of the design in order that its advantages might be practically illustrated," and the device on show at the Juvenile Exhibition was the result of this request. According to the few official papers that survived, this statement was, to say the least, a little misleading. An incomplete letter from Johnson to the secretary of defense dated 10 December 1891 confirmed that he had "now finished the model" and also enclosed some newspaper clippings, presumably those which were found in the archive's file and which were quoted earlier. While there is no direct evidence of an official request to construct a model, Johnson does mention receiving a letter from the Defence Department written on 17 November 1891.

The board appointed to examine Murphy's "Victoria" torpedo apparently also looked at Johnson's device, for they added by way of a postscript to their report of 30 October, "As regards the proposals and sketches of an electrically controlled torpedo from Mr Alfred Johnson of Hamilton, they are in such an elementary and crude state that the Board can form no opinion on them. As far as they go it does not seem they contain any merit."[14] The board's bluntly expressed conclusions were presumably transmitted to Johnson in the letter of 17 November to which the inventor had referred in correspondence.

A letter to Major Cairncross from the Defence Department, which was sent on 19 January 1892, reveals succinctly what the experts thought about the weapon:

Dear Major Cairncross,

I have examined Mr Johnson's torpedo and will write a report to the Defence Department.

So far as I am concerned if you want to get rid of it, I certainly shall not want to see it again.

Yours faithfully

[Signature indecipherable][15]

A note scribbled on the bottom of the letter reads, "Letter sent to Mr Johnson asking him to remove the model 28 January 1892."

It seemed that Johnson's "masterpiece of ingenuity" was unloved and unwanted. Certainly nothing further was heard of it after the Hamilton Juvenile Exhibition. It somehow seemed appropriate that its final appearance was in the guise of an entrancing toy put on display for the sole purpose of amusing the local children.

As the end of the century approached, it was clear that torpedo technology had become increasingly more complicated and was no longer a fruitful field for the amateur inventor. Nevertheless a few hardy souls still tried to break into the lucrative realm of underwater weaponry, although by then most development work had passed into the hands of experienced engineers and those with specialist knowledge of the torpedo. Soon, with rare exceptions, significant advances were only made by anonymous back-room scientists working in the world's great armaments corporations or state-controlled military arsenals with the inevitable consequence that the originators of many brilliant developments in torpedo design remain unrecorded and unknown. In the industrial world of today, teamwork and specialization is the key to success, and there is little room for the many gifted eccentrics who made the nineteenth century such an interesting and rewarding period for historians. The glamour has gone, along with such fascinating personalities as John Ericsson, John Lay, Robert Whitehead, and Louis Brennan. History is the poorer for their loss.

Few details remain of Patrick Cunningham's rocket-propelled torpedo, although, unusually for underwater weapons of this period, a photograph has survived. The latter is dated July 1893 and provides the only clue we have concerning the year of the torpedo's origin. Cunningham

200 · *Nineteenth-Century Torpedoes and Their Inventors*

The Cunningham rocket torpedo and its inventor, New Bedford, Massachusetts, July 1893. The rear end of the launching tube appears at right.

was an immigrant Irish shoemaker living in New Bedford, Massachusetts, and what prompted his interest in a rocket projectile has never been ascertained, although like Philip Holland[16] and his submarine, it may have had a Fenian association. Unlike so many others, the weapon was not only built but was also subjected to a series of tests at the U.S. Navy's Newport torpedo station. However, its short range and uncertain trajectory made it unsuitable for use in naval service, and the weapon was rejected. With typical Irish bravura, whether in protest at the Navy's rebuff or to celebrate some other unconnected event, Cunningham fired the torpedo down the main street of New Bedford. In probably one of the most bizarre episodes in torpedo history, it ended up in a butcher's shop, where it burned down the ice room.

Even less is known of another torpedo thought to have been completed the same year. Except for the name, the Halpine Storage Battery torpedo,

An Eminently Humane Weapon · 201

nothing else is known beyond Bethell's brief mention.[17] As this was clearly one of the antecedents of the modern electric torpedo, it is hoped that a diligent researcher will ultimately chance upon details of its design and history. There is an outside possibility that the storage battery torpedo was developed out of an unusual "exercise" weapon patented by Lt. Nicholas J. Halpine, USN, in the United States in 1891,[18] although, of course, it cannot be certain whether there were two inventors with the same surname or just the one identified in the patent.

The purpose of Halpine's "marine torpedo" was to enable the use of a live warhead during experimental trials without the ensuing loss of the torpedo when it exploded on striking the target. Very briefly, the weapon contained an open-mouthed chamber which carried the explosive charge. The latter was released by a trip mechanism triggered by a harpoon-like device set into the torpedo's nose. The device could be activated mechanically or by electricity, and its layout and external appearance closely resembled the weapon Miles Callender had patented in 1862 with the explosive charge angled at 45 degrees in a lower midship position. Apart from this apparent similarity, however, the two torpedoes were totally different in both purpose and function. In Callender's weapon, the charge was propelled downward by compressed air or rocket combustion and then, by means of a rope or chain pivot, swung forward to strike the target vessel below the waterline. Halpine's declared intention was "to construct a torpedo that just prior to the explosion, the explosive will be delivered from the hull [or torpedo] and the hull [or torpedo] removed from close proximity thereto."

It was quite a neat idea, but despite the promised economy, it was not taken up by the U.S. Navy. Rather surprisingly, another "exercise" torpedo was patented in 1899 by Andrew van Bibber of Cincinnati, Ohio.[19] This was far simpler in concept that Halpine's device and need not detain us for long. It was, in fact, no more than a standard torpedo with the warhead removed from the bows and fitted, instead, on the end of a spar. The torpedo ran on the surface, and only a small charge of explosives was to be carried in the warhead. The intention was to produce a training weapon "of a simple and inexpensive nature." The result was a device that served very little practical purpose.

Australia continued its reputation as a cradle of torpedo invention with Capt. Thomas Wemyss Just's rather extraordinary double-headed

weapon, which was reported on by the local Melbourne press in 1895.[20] This strange concept was a combination of projectile and subaquatic torpedo. Its diameter was given as ten inches and its cigar-shaped body, "tapering to a point at both ends," measured eight feet in length. However, the bow was double-cased with a two-foot-long detachable cap that mirrored the contours of the body shell, leaving the latter with its original tapered shape after the cap had been removed—the only analogy that springs to mind if that of a banana before and after peeling. The only effect of removing the cap was a reduction in its length to six feet.

The weapon was designed to be fired from a special deck gun and was provided with a small propeller and motor which was activated as soon as the main body entered the water. The nature of the engine was not indicated in the report, nor were details provided of any depth-keeping gear which Just might have incorporated into the weapon. The newspaper, however, informed its readers that the body was made, unusually for a torpedo, from aluminum. The thinking behind the weapon was ingenious, although the report leaves a suspicion that the design had not progressed beyond the drawing-board stage. In action it would have been spectacular, especially as its range was claimed to be four miles, far in excess of other contemporary torpedoes.

Having been fired into the air, the bow section and its charge separated from the main body in the manner of a modern two-stage rocket, leaving the latter to follow the former toward the target on a similar but slightly slower trajectory. The bow charge, or cap, being the lighter component proceeded on a slightly higher plane than the main body and was intended to strike the target's upperworks; "a turret . . . may be shattered," the press report opined optimistically. The main body, also carrying a dynamite charge, would enter the water almost simultaneously to inflict further damage below the waterline as it detonated. Tail fins were fitted so that the torpedo's attitude could be adjusted no matter at what angle it entered the water, while a "ventral fin or keel" on the underside of the main body "would ensure steadiness in the water or in the air." The reporter could have added "and pigs might fly," but he refrained from comment.

According to Millar, Captain Just presented his drawings to the Victoria minister of defense with a demand for letters of introduction to the British Admiralty and War Office, but his peremptory request was ignored.[21] The newspaper report, however, states that "the Ministry has given him

credentials, through the Agent-General, to the War office and Admiralty which will be exceedingly valuable."[22] At this distance in time it is not possible to assess which of these two versions was correct. One thing is nevertheless certain: neither the British army nor the Royal Navy saw fit to follow up the idea.

Thomas Just, however, had not given up and took out an American patent in 1899. He had, by then, moved to Brooklyn, New York, and described himself as a "citizen of the United States."[23] The American patent specification showed the projectile weapon to be substantially the same as that already rejected by the Victoria and British governments. It would seem that he fared no better with the U.S. administration, for nothing more was heard of it–but the heading above the preamble of the patent specification reads, "Thomas Wemyss Just, of New York, N.Y., Assignor to Mary Josephine Alsbau, of same place," which suggests that he had at last found a gullible buyer for his paper project even if the world's governments were not interested.

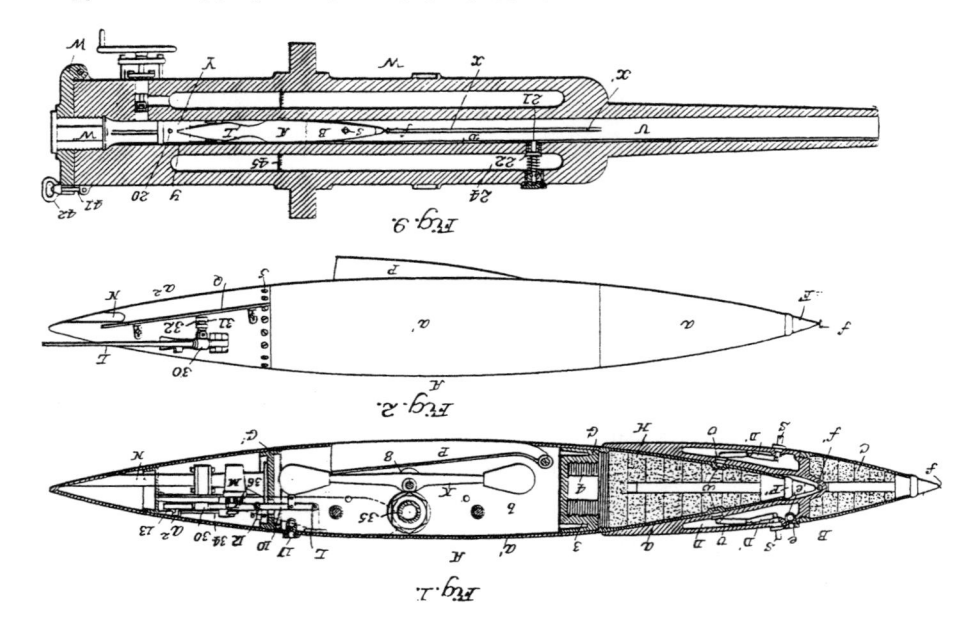

Thomas Wemyss Just's unusual double-headed torpedo was fired from a gun like a projectile and separated in midair so that it could strike the target's upperworks and explode below the waterline at the same time.

The press report on the Just weapon contained an alarming insight into the reality of the contemporary torpedo's efficiency. For having observed that the Whitehead was regarded as "about the most perfect type in use," it proceeded to review a test carried out at Woolwich earlier that year (1895): "Of eight [torpedoes] which were fired three failed altogether to run; three others ran but failed to hit anything; one hit something it was not aimed at . . . and only one struck its intended mark."[24] Even the weapon that hit the wrong target did it "so gently as to render it doubtful whether it would have exploded." Perhaps it was fortunate that gyroscopic steering control was introduced later that year.

New ground was broken in 1896 with a magnetic torpedo designed by Louis F. Johnson, Walter J. Slacke, and Howard Lacey, all of whom were residents of Easton, Pennsylvania.[25] This invention was based on the fact that "while it is not difficult to bring the torpedo into proximity to the vessel . . . to be destroyed, it is sometimes difficult to bring it into contact therewith." Their solution was to fit an electromagnet into the bows of the weapon "which will attract it to be held against the iron or steel vessel." It was not a totally new torpedo as such but a modification intended for the Whitehead and other conventional locomotive weapons. Still, it showed some new thinking, something that was greatly needed by the time the end of the century approached. The magnet was shaped to fit into the contours of the torpedo's nose—so snugly, in fact, that the drawings show no space available for the warhead. Sadly, the three inventors appear to have overlooked this rather important flaw in their idea, for as they were proposing to use a standard torpedo, there would have been no other spaces available to accommodate the explosive charge, the interior of the locomotive torpedo being as tightly packed as an overstuffed Thanksgiving turkey.

George W. McMullen of Chicago, Illinois, also employed magnets in the weapon he had invented sixteen years earlier in 1880.[26] But this was more in the way of a self-propelled mine than a legitimate torpedo, for it was intended to settle on the sea bed and wait for a likely target to approach. The magnets, operated electrically by an observer on shore, were used to release floats and drop the keel of the device so that it would rise toward the target. The patent, however, gave no indication of how the explosive charge, which was omitted from the drawing, was to be triggered, but it must be assumed that this would have been accomplished electri-

cally by the onshore operator. McMullen's weapon was really no more than an observation mine and has only been included to show that other, inventors had given thought to the use of magnets in the nineteenth century. It was really not until World War II that magnetism began to flourish as one more means of sinking enemy ships. But that, of course, is another story.

A completely new weapon was proposed by Thomas E. Barrow of Mansfield, Ohio, which, according to his specification,[27] could be sent through the water in any direction without being fired from a torpedo tube. Additionally, it could be triggered not only by impact but also by means of an internal electrical time-delay mechanism. This latter statement, which formed part of the patent specification, is, however, slightly misleading, for by line fifty Barrow was referring to a train of gears "to be operated with a clock-spring for exploding the . . . charge at a given time." One can almost picture the ghostly hand of Robert Fulton guiding Barrow's pen as he wrote this section. For some strange reason the specification failed to mention the important fact that the torpedo was to be propelled by electricity supplied from a storage battery, a power source that placed Barrow's torpedo firmly into the twentieth century.

A curious feature of the design kept the "torpedo shell," or warhead, separate and distinct from the outer casing of the main body. The battery, too, was stored independently, and the various elements were only brought together to make up the complete weapon shortly before use. It is difficult to understand the purpose of this separation—or, indeed, its advantages. And Barrow failed to give any reasons for adopting this strange arrangement. It can only be concluded that the body was no more than a launching vehicle for an explosive charge. The result is a puzzling cross between Miles Callender's concept of 1862, which seems to have inspired several other inventors, and an unmanned miniature submarine from which the "torpedo" was to be released. Not surprisingly, the device never got further than the drawing board because it was thoroughly impractical, not least because the inventor had taken little or no account of how the contrivance was to maintain depth.

Hudson Maxim reappeared in the torpedo story in 1898, when he sought patent protection for an improved automobile torpedo.[28] Having observed that in order to generate enough power "it has . . . been found necessary to construct a reservoir of such size as to occupy the greater

proportion of the space within the torpedo, thereby greatly restricting the space to be utilized in carrying the explosive charge," Maxim sought to overcome this drawback by devising a means of limiting the space taken up by the propellant–air, gas, or liquid–by using an explosive additive to boost the force of the conventional propellant. He also recognized that this would heat the air so that "it will not freeze or produce appreciable cold, and thus hinder the proper operation of the motor."

His specification proposed that a suitable combustible nitro-compound should be burned inside the torpedo's shell "in such a manner that the products of [the] combustion may combine with or enter with the motive fluid into the torpedo motor, so that not only is the pressure of the motive fluid increased or maintained for a longer period of time and its volume augmented, but the air is so heated that it will not freeze or produce appreciable cold, and thus hinder the proper operation of the motor." Earlier in the specification Maxim had observed shrewdly that heating the propelling medium would produce higher speeds.

It is not possible to examine the invention in detail as this book does not purport to be a technical work. But in the broadest terms, Maxim suggested a form of open combustion (contained within a small chamber) with a pipe leading to the motor where the resultant heat could be combined with the propelling gas or liquid to create a more potent mixture. As repeatedly noted throughout this account of torpedo development, the heater principle had been recognized and exploited from von Scheliha's "furnace" to Lay's "naked flame" plus those relying solely on the warming effect of seawater. So Hudson Maxim's heater was really nothing new. Unfortunately, it arrived on the scene just a few months too late for him to reap any reward. Within a year the Royal Gun Factory at Woolwich had finally awakened to the benefits of the heater concept and was already feverishly working on designing an appropriate system–quickly followed by the Armstrong and Whitehead factories. Backed by government money and unlimited private finance, against which Maxim had no hope of competing, his ideas were swept aside in the rush to develop the most efficient and effective means of heating compressed air and obtaining higher performance.

But the 1900 patent had already given Hudson Maxim a head start in the increasingly important field of torpedo fuels, for in addition to the heater, it also contained the seeds of his explosive propellant Motorite,

which he patented in more precise detail in 1904.[29] This new insight into fuel technology helped to confirm his reputation as one of the world's leading experts in the field of torpedo propellants. Although these later events fall outside the strict parameters of this narrative, another patent, granted in 1921 after his death, needs to be mentioned.[30] This covered "a liquid explosive fuel for driving torpedoes [and] consisting of a nitro-compound of glycerine holding in solution gum camphor, acetylene gas, acetone and a combustible non-explosive material [for which] it was proposed to employ a tungsten generation or combustion pot."[31] It would be true to say that today's propellant technology, which is now a branch of science in its own right, has evolved from Hudson Maxim's pioneering work in the closing years of the nineteenth century.

Although most of Frank McDowell Leavitt's patents were taken out after 1900, it is impossible to close this account of torpedo development without considering his contribution to the weapon's history—if only for the part he played in introducing the Whitehead to the U.S. Navy. He was undoubtedly the most important and influential figure in the field of American torpedo production during the first two decades of the twentieth century.

Born in Athens, Ohio, he studied at the Stevens Institute of Technology at Hoboken, New Jersey, from which he graduated 1875. His first job found him engaged on the design of steam steering gear for the U.S. Navy with Frederick E. Sickels of New York before moving on to Bliss and Williams of Brooklyn as their chief draftsman, a company which was, at that time, manufacturing machinery for the production of sheet metal. Then, in search of fresh opportunities, he joined the Mexican National Railway Company as a master mechanic before transferring his loyalties to the Graydon and Denton Manufacturing Company of Jersey City a year later. Finally, in 1884, he returned to Bliss and Williams, by then known as E. W. Bliss and Company, as assistant superintendent.

Leavitt was recognized by his peers as an engineer of considerable talent not least for his ability to visualize a problem and resolve it with design drawings of complex detail yet expressed in the simplest of terms. Between 1875 and 1921, he was granted over 300 patents, although this number is dwarfed by the astonishing 1,033 taken out by Thomas Edison in his lifetime. Initially engaged on the invention and design of hydraulic presses and dies for use with the company's main line of business of

sheet-metal work, Leavitt quickly rose to become the firm's superintendent and was responsible for installing the machinery in another of Eliphalet Williams Bliss's[32] factories, the United States Projectile Company.

It was during an extended tour of Europe, following his appointment as
E. W. Bliss and Company's chief engineer, that he visited the Whitehead
factory in Fiume and was able to examine the torpedo that was, by that
time, universally recognized as the leader in the field of self-propelled
underwater weapons. And it had not escaped his notice that for some un-
accountable reason the U.S. Navy possessed not one single example of
Whitehead's torpedo. Indeed, apart from the home-grown Howell, it was
showing virtually no interest in the weapon that was to dominate combat
at sea in two world wars.[33] The various contracts that Robert Whitehead
negotiated from 1868 onward have never been located and their details
remain unknown. For this reason the background of the agreement with
Leavitt for the manufacture of the Fiume torpedo under license in the
United States is shrouded in mystery and it is impossible to say whether
Whitehead or Leavitt first proposed the deal. External evidence suggests
that the contract was signed in 1891. Its terms were unprecedented, for it
was the only occasion Robert Whitehead entrusted the construction of
his weapon to a private company rather than a government. Indeed, the
relevant column in the factory's ledgers was headed "Government" and
showed only the country to which the delivery was consigned. Knowing
the business acumen of both Whitehead and his partner, Count Georg
Hoyos, it is probable they realized that this was the only way they could
penetrate the lucrative American market which had proved so curiously
resistant to the pressures of the Fiume-based manufacturer. Nevertheless, even they were unaware of the golden goose they were about to en-
snare, for in 1891, the United States was by no means the world-ranking
naval power it was to become in the course of the ensuing twenty years.

Under the terms of the contract to which Leavitt had appended his sig-
nature, E. W. Bliss and Company was to be provided with all the necessary working drawings together with the relevant sample torpedoes.
Confident in his own expertise to produce the machines and tools that
were needed to manufacture the weapons, and conscious of the resulting
cost savings, he did not, however, purchase any of the Fiume company's
machinery. This proved a wise decision, for, with typical flair, he was
able to design and build all the requisite tools and plant on Bliss and

An Eminently Humane Weapon · 209

Company's Brooklyn premises in readiness for volume production. Yet despite the enormity of this self-imposed task, Leavitt continued with his usual duties on the sheet-metal side of the business until 1900, although after that date, most of his time was devoted to torpedo development.

The Fiume factory's consignment ledger actually pinpoints the two sample torpedoes delivered to Bliss and Company under the terms of the licensing agreement, although the entry for the consignee appears in the "Government" column and is shown merely as "United States."[34] But this is a misleading error by the bookkeeper, who had never before been required to record a private transaction. The first weapon, a 14-inch model, serial number 4243, was consigned on 13 May 1891 and was followed, on 1 July 1891, by a 45-centimeter (17.7-inch) model with the serial number 4244. The pre-delivery test performance of the latter was noted as 28.4 knots over eight hundred yards. This weapon would have been a Fiume 18-inch Mark II, weighing 1,136 pounds and carrying a 187-pound warhead. With an overall length of sixteen feet seven inches, and powered by a three-cylinder Brotherhood radial engine, it was an altogether faster and heavier weapon than the 14-inch delivered to Brooklyn six weeks earlier.

Having evaluated the two patterns, and no doubt discussed their respective merits with members of the Board of Ordnance, F. W. Bliss and Company went ahead with the 45-centimeter (more usually known as the 18-inch) prototype, which thus became the first Whitehead-based torpedo to enter service with the U.S. Navy. It took time, however, for production to get seriously under way at the Brooklyn factory, and Bethell noted that there were still "fewer than 200 Whitehead torpedoes" in service in 1896 (other sources dispute this).[35] In fact, various figures for the U.S. Navy's stock of Whitehead/Bliss torpedoes have been bandied about by differing experts and without official figures–to which access appears to be denied–it is probably wiser not to place much credence on the total number said to be in service during the period from 1894 to around 1907.

Some authorities state that all-American versions of the Fiume weapon were classified as Bliss-Leavitts by the U.S. Navy, but internal records show that this was not necessarily the case. An official U.S. Navy manual issued in 1914 uses only initials in the column headed "Type."[36] And while the later 18-inch and all 21-inch weapons are noted as *BL*, pre-

sumably for Bliss-Leavitt, the initial batches of 18-inch Marks I, II, and III are entered as *WH,* which could stand either for "Whitehead" or, possibly, "Whitehead Heater." As these three patterns were shown as obsolete by 1914, they were presumably the original three models delivered by Bliss and Company in and after 1894.

A further diversion into the twentieth century is necessary to clear up another minor puzzle concerning the U.S. Navy's Whitehead torpedoes. An article in the *Submarine Review* observed that an "often overlooked Whitehead torpedo was used by the U.S. Navy at this time [not later than 1906?], namely the Whitehead 5-metre Mk 1A which was purchased directly from Whitehead." A footnote suggested that the weapons in question were probably the Fiume 18-inch Mark I or Mark II, and the main text adds "fewer than fifty 5-metre Mk 1A torpedoes were purchased."[37]

The Whitehead factory ledger certainly corroborates this rather odd arrangement, which, it must be assumed, was entered into with the full knowledge and agreement of the American licensees, Bliss and Company. But delivery was much later than the dates implied in the article, the ledger entry showing 31 July 1913. This confirms the dispatch of five 45-centimeter (17.7-inch) weapons, serial numbers 12696–12700, and the consignment also included two of the new 53-centimeter (usually known as 21-inch) torpedoes, each capable of carrying a warhead charge of 180 kilograms (396.8 pounds). These last two weapons had the serial numbers 12701 and 12702. Significantly, all seven were noted in the ledger as having two-cylinder horizontal engines, the first "wet heater" power unit to be produced by the Fiume factory. No other direct-sale Whiteheads were listed as supplied to the U.S. government, and it would appear that this small batch was obtained from Fiume so that the U.S. Navy's Bureau of Ordnance could appraise the benefits of Whitehead's new heater system. Why the Navy took it upon itself to obtain direct delivery of these new heater models instead of working through Bliss and Company has never been revealed.

It was, incidentally, the size of the 21-inch weapon's warhead that helps to explain why the Howell flywheel torpedo was discarded so quickly, for even at an early stage, the bureau had clearly recognized the development potential of Bliss and Company's initial batch of 18-inch Whiteheads.

The first Bliss-Leavitt to incorporate the new heater technology was the 18-inch Mark VII, built around 1907, the same year Fiume also produced its first dry-heater weapon. Woolwich arsenal's Royal Gun Factory (RGF), however, did not complete its own heater-engined torpedo until 1909–this being the RGF 18-inch Mark VII (H). It is clear that Bliss was keeping level with Whitehead's factory and ahead of British designers so far as the improved engines were concerned. But, perhaps, as licensees, Frank Leavitt and his team enjoyed a little extra help and guidance from Fiume.

Remaining in the twentieth century, much has been made of Leavitt's first Curtis turbine-engined torpedo which appeared in 1905 as the 21-inch Mark I, although development work had been ongoing since 1903. Woolwich, however, had already experimented with a Parsons's cold-air turbine, which was fitted into a 14-inch weapon, between 1897 and 1899, but as its performance proved inferior to the equivalent Brotherhood compressed-air power-unit, additional work on the idea was abandoned. In any event, the rediscovery of the heater principle soon directed the minds of those seeking improved performance in a new and more exciting direction. In fact, and not unexpectedly, the first in the field had been Robert Whitehead. For in a letter he wrote in 1891,[38] he referred to experiments with a small turbine, based on the Dow system, "which promises well," although he admitted to difficulties in gearing down its high rotational velocity of up to forty thousand rpm. As the power plant was never accorded a public debut, it must be assumed that Whitehead abandoned the idea–thus leaving Frank Leavitt to produce the first workable turbine-engined torpedo in 1906 or 1907.

These developments have only been described in outline as they mainly occurred after the 1900 watershed–although several, as we have seen, had their origins in the nineteenth century. Leavitt himself was a nineteenth-century inventor, although, so far as the torpedo was concerned, he did not really blossom until after 1900, when he concentrated his attention on the weapon by which he is now best remembered. Yet despite his expertise, he was to spend part of the Great War chairing the Committee on Experimental Power for the Navy Department's Bureau of Steam Engineering. And believe it or not, he was mainly involved in the development of a steam-powered aircraft! With the torpedo proving itself

to be the most important weapon in the war at sea, especially in the hands of the Kaiser's U-boat commanders, it was a scandalous waste of talent and experience. Leavitt died on 6 August 1928.

Before closing this account of nineteenth-century torpedoes and their inventors, it is time to mention what was obviously the most remarkable underwater weapon of the period. Although not reported in the *Engineer* until 1901, it was said to have been tested "before the King, in Sweden, a couple of years or more ago," which would place it firmly in the century under review. Referred to only as the Orling torpedo, presumably based on the name of its inventor, the periodical alleged that "there is . . . more than one torpedo in the field worked with wireless telegraphy, X-rays, or some similar force, as [its] steering agency. . . . Something of the same kind is supposed to exist in the torpedo school ship *Vernon*. From Italy similar stories have come.[39] All told, this type of weapon exists today in a more or less perfected form."[40]

Leaving aside the more fanciful aspects of the report, such as the employment of X-rays, which the writer seems to have confused with radio waves, and making due allowance for journalistic hyperbole, the guidance system (assuming that the Orling did exist) seems most likely to have been based on a rudimentary system of wireless telegraphy. The reception of radio waves was first demonstrated by Heinrich Hertz in 1887, and the first practical application of his research came from the English scientist Oliver Lodge in 1894. This neatly places wireless telegraphy in the appropriate time frame for its adoption as a means of remotely steering a dirigible torpedo in, say, 1897 or 1898. And while Victorian and Edwardian science fiction writers were fascinated by X-rays and their more dramatically exciting extension, death rays, it seems logical to conclude that Orling had produced some form of wireless-controlled guidance apparatus which, as no more was heard of it, presumably failed to work.

However, odd comments in the report are of interest insofar as they reflect contemporary thinking on torpedo warfare at the turn of the century. "The torpedo," opined the writer, "is an eminently humane weapon, the odds being that a big ship thus struck will not sink but be merely *hors de combat*." One wonders whether he lived long enough to witness the destruction of the battleship *Barham* after it had been struck by a salvo of three torpedoes from von Tiesenhausen's *U-331* on 25 November 1941 off the North African coast.

But the journalist's concluding sentence sums up the very essence of this narrative: "A rather large gap usually exists between the real and the ideal."[41] Having come this far, and bearing in mind some of the weird and wonderful weapons we have considered on our journey, the reader will no doubt agree with his sentiments.

Appendix *British and U.S. Patents*

The following patents are courtesy of the British Library, London.

British Patents

7104	31 May 1836	F. P. Smith
7149	13 July 1836	J. Ericsson
114	15 January 1861	R. Wilson
3536	25 November 1872	V. von Scheliha
1585	7 October 1873	V. von Scheliha
4069	23 November 1875	V. von Scheliha
2584	23 June 1876	R. Wilson
3359	4 September 1877	L. Brennan (Backdated from 1878)
5363	21 December 1880	J. Ericsson
2614	25 May 1883	W. Sims
4494	20 September 1883	T. Nordenfelt
4776	8 October 1883	Haight, Wood and Winsor
16693	1888	T. Nordenfelt
10581	13 September 1890	Ericsson Coast Defence Company
9774	24 June 1891	G. R. Murphy
12039	2 June 1896	G. Marconi

United States Patents

35285	13 March 1862	W. H. Elliot
35957	22 July 1862	O. C. Smith
36654	14 October 1862	I. A. Ketcham
39612	16 October 1862	M. L. Callender (antedated)
37942	17 March 1863	B. F. Smith Jr.
41112	5 January 1864	J. D. Willoughby
46850	14 March 1865	J. L. Lay
46851	14 March 1865	Wood and Lay

46852	14 March 1865	Wood and Lay
46853	14 March 1865	Wood and Lay
48124	6 June 1865	J. D. Willoughby
49300	8 August 1865	G. M. Ramsey
58661	9 October 1866	S. S. Merriam
121052	21 November 1871	J. A. Howell
134493	31 December 1872	H. J. Smith
156320	27 October 1874	C. J. von Court, Exors of
158501	5 January 1875	H. F. Knapp
199841	29 January 1878	W. H. Mallory
202453	16 April 1878	H. Mortensen
211301	14 January 1879	J. L. Lay
211302	14 January 1879	J. L. Lay
211303	14 January 1879	J. L. Lay
219711	16 September 1879	W. Giese
222718	16 December 1879	J. H. McLean
223855	27 January 1880	W. H. Mallory
227637	18 May 1880	G. W. McMullen
245864	16 August 1881	G. H. Reynolds
246415	30 August 1881	G. H. Reynolds
249192	8 November 1881	W. H. Mallory
250144	29 November 1881	G. E. Haight
257693	9 May 1882	G. E. Haight
257694	9 May 1882	G. E. Haight
273413	6 March 1883	A. Weeks
292428	22 January 1884	Haight, Wood, and Winsor
311325	27 January 1885	J. A. Howell
312579	17 February 1885	T. Nordenfelt
319633	9 June 1885	W. S. Sims
327380	29 September 1885	W. I. Chambers
339096	30 March 1886	Haight and Wood
358471	1 March 1887	S. H. Nealy
359313	15 March 1887	H. M. Bennett
364364	7 June 1887	A. P. S. Miller
370570	27 September 1887	Sullivan and Etheridge
399516	12 March 1889	R. J. Gatling
420406	28 January 1890	W. K. Cavett
431210	1 July 1890	V. F. Lässoe
442327	9 December 1890	G. R. Murphy

450875	21 April 1891	W. S. Sims
453861	9 June 1891	N. J. Halpine
458677	1 September 1891	J. A. Howell
478215	5 July 1892	H. Berdan
559711	5 May 1896	Johnson, Slacke, and Lacy
613809	8 November 1898	N. Tesla
627312	20 June 1899	A. van Bibber
632089	29 August 1899	T. E. Barrow
638463	5 December 1899	T. W. Just
641787	23 January 1900	Hudson Maxim
1683085	23 February 1921	Name unknown

Notes

Unless already identified in the notes, full details of publishers and publication dates of all books referred to below will be found in the bibliography.

Introduction

1. The story of Whitehead's life and the development of his torpedo, together with more detailed accounts of earlier underwater weapons, will be found in my *Devil's Device*.

Chapter 1. Defective in Principle

1. Gray, *Devil's Device*. Page references refer to the enlarged second edition published by the Naval Institute Press in 1991.

2. Caseli and Cattaruzza, *Sotto i mari del mondo*. In 1868, Whitehead negotiated a new partnership with de Luppis which gave the latter a share of the profits from the initial sales of the torpedo but left Whitehead free to develop the weapon autonomously and to retain the profits from this subsequent work. De Luppis died in Milan in 1875 "embittered and disappointed" over losing control of "his" weapon.

3. Public Record Office, London, ADM 116/163. Hereafter cited as PRO.

4. *Engineer*, 15 December 1871.

5. Gray, *Devils Device*, 57–58.

6. The history of the spar torpedo will be found in chapter 6, and the history of the Harvey towed torpedo will be found in chapter 7; *Engineer*, 1 December 1871.

7. Admiral Lord Fisher, *Records*. A detailed account of Fisher's involvement with the torpedo will be found in Gray, *Devil's Device*.

8. PRO ADM 116/135.

9. *Engineer*, 15 December 1871.

10. Ibid.

[219]

220 · *Notes to Pages 16–28*

11. British Patent Office No. 1869, 27 July 1864. (Similar references hereafter cited as "British Patent.")

12. *Engineer*, 15 December 1871.

13. Ibid.

14. Ibid.

15. *Times* (London), 11 November 1870 (subsequent references to the *Times* are to the *Times* of London).

16. *Engineering*, 30 December 1870.

17. *Engineer*, 10 January 1873.

18. *Engineering*, 1 April 1870, 213.

19. PRO ADM 116/163, p. xi of the report.

20. Gray, *Devil's Device*, 46–48.

21. Church, *Life of John Ericsson* 2:160.

22. Letter from Col. G. Frazer of the Royal Laboratory, Woolwich, 23 June 1877, quoted in Wilson, *Who Invented It?*

23. Ibid.

24. British Patent No. 7149, 13 July 1836.

25. Bradford, *Notes on Movable Torpedoes.*

26. Kirby, "Development of Rocket-propelled Torpedoes," 2; and Kirby, "History of the Torpedo," in vol. 27, no. 1, p. 37.

27. U.S. Patent No. 39,612, 16 October 1862.

28. Milford, "U.S. Navy Torpedoes."

29. U.S. Patent No. 121,052, 21 November 1871.

30. *Engineer*, 17 August 1888.

31. Bethell, "Development of the Torpedo," 3:302, 10 October 1945.

32. U.S. Patent No. 311,325, 27 January 1885.

33. Bethell, "Development of the Torpedo," 3:302, 10 October 1945.

34. Gray, *Devil's Device*, 57–58.

35. Ibid., 145 and 152, passim.

Chapter 2. The St. Petersburg Connection

1. A commendable exception was Rear Adm. E. N. Poland's *Torpedomen*, which provides considerable details of the committee's work. My own source was the official report, issued in July 1876, filed at the PRO under ADM 116/163.

2. Gray, *Devil's Device*. Chapter 1 is devoted to the 1870 trials.

3. Poland, *Torpedomen*, 22.

4. Admiralty Torpedo Committee, *Final Report*, x; emphasis added.

5. Admiralty Torpedo Committee, *Final Report*, quoted in Poland, *Torpedomen*, 24.

Notes to Pages 28–42 · 221

6. British Patent No. 3536, 25 November 1872.

7. Biographical material supplied by the Hartley Library of Southampton University by courtesy of C. M. Woolgar, head of Special Collections.

8. Wellesley, *Recollections*, 122.

9. Ibid.

10. Ibid.

11. British Patent No. 3536, 25 November 1872.

12. Dates extracted from the Whitehead Company's consignment ledger.

13. Gray, *Devil's Device*, 89–90.

14. Ibid., 156; emphasis added.

15. Ibid., 179.

16. See chapter 8.

17. British Patent No. 1585, 2 May 1873 (sealed 7 October 1873).

18. British Patent No. 4069, 23 November 1875.

19. *Times*, 27 October 1874.

20. *Times*, 6 October 1874.

21. Wellesley, *Recollections*, 122.

22. Ivan Alexandrovsky was also claimed to be the inventor of the submarine, although no firm evidence has been produced to confirm the assertion.

23. See chapter 6.

24. Vice Admiral Popov was responsible for the construction of two circular ironclads built for the Russian Navy in 1875. Although based on the design of Ericsson's *Monitor*, they proved unsuccessful.

25. Caseli and Cattaruzza, *Sotto i mari del mondo*. The date given differs from that in *Devil's Device* (October 1868) but is probably more accurate as the authors had access to the company's archives.

26. This brief account of pioneer Russian torpedoes has been gleaned from various printed sources published in Russia and Belarus. The text has been privately translated. While they are believed to be reliable, I am unable to vouch for their complete accuracy. As the original texts were supplied in extract and manuscript form, it has not been possible to identify the titles of the books or periodicals in which they appeared.

Chapter 3. Extreme Simplicity

1. Bethell, "Development of the Torpedo," 1:442.

2. U.S. Patent No. 46,850, 14 March 1865.

3. Bethell, "Development of the Torpedo," 1:442.

4. Ibid.

222 · *Notes to Pages 42–55*

5. U.S. Patent Nos. 46,850, 46,851, 46,852, and 46,853, all dated 14 March 1865.

6 U.S. Patent No. 46,853.

7. Bethell, "Development of the Torpedo," 1:442.

8 Admiralty Torpedo Committee, *Final Report*, xi.

9. *Engineering*, 7 February 1883, 107.

10. Admiralty Torpedo Committee, *Final Report*, xi.

11 *Engineering*, 7 March 1873, 175, fig. 2.

12. Admiralty Torpedo Committee, *Final Report*, xi.

13. Cdr. William A. Kirkland's letter to the *Army & Navy Journal* (n.d.), quoted by Ericsson in *Engineering*, 7 March 1873, 107.

14. *Engineering*, 7 February 1873, 107, and 7 March 1873, 175.

15. Ibid.

16. Gray, *Devil's Device*, 156.

17. *Engineering*, 7 February 1873, 108.

18. Ibid.

19. Admiralty Torpedo Committee, *Final Report*, xi.

20. Gray, *Devil's Device*, 93.

21. *Engineer*, 2 April 1880, 243–44.

22. *Engineering*, 7 February 1873, 107, figs. 1 and 2.

23. *Engineer*, 2 April 1880, 243.

24. Ibid.

25. Ibid.

26. Ibid.

27. U.S. Patent Nos. 211301 and 211302, both dated 14 January 1879.

28. Preamble to U.S. Patent No. 211302. Ibid.

29. *Engineer*, 2 April 1880, 243.

30. *Engineering*, 7 March 1873, 107.

31. British Patent No. 3536, 25 November 1872.

32. *Engineer*, 2 April 1880, 243.

33. Clowes, *Four Modern Naval Campaigns.*

34. Ibid.

35. Gray, *Devil's Device*, 106–9.

36. Hobart Pasha, a Royal Navy officer, was de facto head of the Ottoman Navy.

37. *Times*, 27 December 1882.

38. Ibid.

39. Ibid.

40. Bethell, "Development of the Torpedo," 7:243.

Notes to Pages 55–65 · 223

41. *Engineer,* 2 April 1880.

42. *Times,* 2 October 1883.

43. *Engineer,* 11 March 1887, 191.

44. Bethell, "Development of the Torpedo," 3:301.

45. Kirby, "History of the Torpedo," in vol. 27, pt. 1, p. 38.

46. U.S. Patent No. 292,428, 22 January 1884.

47. Kirby, "History of the Torpedo," in vol. 27, pt. 1, p. 38, fig. 9.

48. Sleeman, *Torpedoes and Torpedo Warfare.*

49. Sleeman, "Torpedoes," 495.

50. Ibid.

51. Ibid.

52. Ibid.

53. U.S. Patent No. 292,428, dated 22 January 1884, showed the following foreign patents granted: France, 25 September 1883; Belgium, 25 September 1883; Germany, 26 September 1883; Spain, 28 September 1883; Italy, 29 September 1883; and England (No. 4776), 8 October 1883.

54. U.S. Patent No. 158,501, 23 September 1873.

55. *Engineer,* 11 March 1887, 191.

56. Sleeman, "Torpedoes," 495.

57. *Engineer,* 11 March 1887, 191.

58. Ibid.

59. Ibid.

60. Sleeman, "Torpedoes," 495.

61. *New York Times,* 21 April 1899, obituary.

Chapter 4. A Highly Respected Scottish Engineer

1. Sleeman, *Torpedoes and Torpedo Warfare,* 1889 ed., 179.

2. Bethell, "Development of the Torpedo," 2:5.

3. Kirby, "History of the Torpedo," in vol. 27, pt. 1, p. 34.

4. Gray, *Devil's Device,* 88–89.

5. James Nasmyth (1808–1890), a famous British engineer and the inventor of the steam hammer.

6. *New York Times,* 9 March 1889, obituary.

7. Gray, *Devil's Device,* 159.

8. Ibid., 25.

9. Church, *Life of John Ericsson* 1:85. Most of Ericsson's biographical details are based on material to be found in this work.

10. Letter quoted in ibid. 1:87.

11. British Patent No. 7149, 13 July 1836.

12. British Patent No. 7104, 31 May 1836.

13. Church, *Life of John Ericsson* 1:97.

14. Ibid.

15. Ibid. 1:96.

16. Bourne, *Treatise on the Screw Propeller,* quoted in ibid. 1:97.

17. Ibid. 1:98.

18. Wilson, *Who Invented It?* This monograph is based on a pamphlet which Wilson had first written in 1860.

19. *Warship International,* 113/68.

20. British Patent No. 7149, 13 July 1836.

21. Sleeman, "Torpedoes," 496.

22. British Patent No. 7104, 31 May 1836.

23. The biographical details up to 1860 are taken from Wilson, *Who Invented It?*

24. Ibid., 8.

25. *Dictionary of American Biography,* 1921 ed. (hereafter cited as *DAB*), 596.

26. Wilson, *Who Invented It?* 13.

27. Ibid.

28. Ibid.

29. Ibid., 17.

30. Gray, *Devil's Device,* 89.

31. *Royal Navy Torpedo Manual,* 1887 (hereafter cited as *Torpedo Manual, 1887*), pl. 1, RL 14-inch Mark I.

32. Wilson, *Who Invented It?* 22.

33. Ibid., 23.

34. Ibid.

35. Ibid.

36. Press report quoted in ibid.

37. Wilson, *Who Invented It?* 25.

38. Ibid., 26–28. Wilson's narrative is not always clear, but he seems to be referring to the clockwork model which he had built for the 1827 tests. It is depicted in figures 4 and 5 on page 19 of his monograph and is described as Model of 1827 and 1828, made in 1826, about three feet long.

39. Ibid., 27.

40. Ibid., 35.

41. Ibid., 36.

42. Ibid., 36 n.

43. Ibid., 37.

44. Ibid.

Notes to Pages 75–91 · 225

45. Ibid., 44.

46. Ibid., 41.

47. Ibid., 17 and 40.

48. Church, *Life of John Ericsson* 1:89.

49. Ibid. 1:95. The *Robert F. Stockton* was surprisingly small for a vessel that was to cross the Atlantic under her own power, even though it took her forty-six days. Built by John and Macgregor Laird of Birkenhead with an iron hull and a 50-horsepower engine, her deck length was only seventy feet and her beam ten feet. Despite her ocean crossing, her draught was just three feet and she was designed for work as a canal tug.

50. *DAB*, 1921 ed., 596.

51. British Patent No. 2584, 23 June 1876.

52. *Torpedo Manual* (1887), 18, subheading: *Wheel gearing.*

53. Wilson, *Who Invented It?* 44.

54. Ibid., 45 n.

55. Church, *Life of John Ericsson* 1:154; emphasis added.

56. PRO ADM 116/164.

57. Delivery of the 14-inch Model A (Royal Navy identification: Fiume 14-inch Mark I) commenced on 26 May 1877, according to the Whitehead company's consignment ledger. The serial numbers of the batch were 361 to 560, inclusive.

58. PRO ADM 116/164.

59. *Times,* 29 March 1873, p. 5.

60. Gray, *Devil's Device,* 55, bottom illustration.

61. Wilson, *Who Invented It?* Appendix B, p. 59.

62. *Torpedo Manual,* 1887, 18.

Chapter 5. Most Splendid Results

1. Bethell, "Development of the Torpedo," 5:530.

2. *Times,* 7 January 1874.

3. Presumably William Froude (1810–79), who built the first ship model testing tank for the British Admiralty.

4. *Times,* 4 May 1878.

5. U.S. Patent No. 222718, 16 December 1879.

6. U.S. Patent No. 219711, 16 September 1879.

7. *Engineering,* 1 April 1870, 213.

8. *Engineering,* 15 April 1870, 257.

9. Ibid.

10. Letter from Ericsson dated 10 April 1874, quoted in Church, *Life of John Ericsson* 2:161.

11. Ibid.

12. Gray, *Devil's Device*, 54, 141.

13. Letter from Ericsson quoted in Church, *Life of John Ericsson* 2:163.

14. Letter from Commodore Jeffers, quoted in ibid. 2:164–65.

15. Ibid. 2:167.

16. Ibid. 2:168.

17. *Engineering*, 15 April 1870, 257.

18. Bethell, "Development of the Torpedo," 7:244.

19. Ibid.

20. *Torpedo Manual*, 1929, para. 51, p. 11.

21. Church, *Life of John Ericsson* 2:173.

22. Ericsson's reply to Jeffers's letter of 7 April 1878, quoted in ibid. 2:165.

23. Telegram dated 7 February 1879, quoted in ibid. 2:172 n.

24. Ibid. 2:175–76.

25. Sleeman, "Torpedoes," 489.

26. Kirby, "Development of Rocket-propelled Torpedoes."

27. Kirby, "History of the Torpedo," in vol. 27, pt. 1, p. 39.

28. *Torpedo Manual*, 1929, para. 54.

29. Bethell, "Development of the Torpedo," 5:530. Also Kirby, "Development of Rocket-propelled Torpedoes," 2 and 3.

30. Kirby, "Development of Rocket-propelled Torpedoes," 2 and 3.

31. Roy M. Marcot's biography of Hiram Berdan (see the bibliography) has assisted in confirming particular aspects of his career and inventions which had already been assembled from a variety of sources. It also proved useful in pinpointing several important contemporary archival sources for further and more torpedo-focused examination.

32. *New York Herald*, 6 June 1886, p. 313.

33. *Times*, 20 April 1882, p. 5.

34. *Times*, 27 December 1882, p. 5.

35. *Times*, 13 June 1882.

36. *General Information Series No. 6*, Office of Naval Intelligence, June 1887, quoted in Warship International 113/68.

37. *Engineering*, 17 February 1871, 120. See also U.S. Patent No. 39612, antedated 16 October 1862.

38. Bethell, "Development of the Torpedo," 5:530.

39. U.S. Patent No. 273413, 6 March 1883.

40. *Torpedo Manual*, 1929, para. 52, p. 11.

41. *Engineer*, 17 October 1884, 293.

Chapter 6. Filled with Cork and Glue

1. Scharf, *Confederate States Navy*, 754.

2. Ibid.

3. Bethell, "Development of the Torpedo," 1:442.

4. *Southern Historical Society Papers* 4, no. 5 (November 1877).

5. Gray, *Few Survived*, 28–34.

6. *Pensacola News Journal*, 7 March 2001.

7. Wilson, *Ironclads in Action* 1:110–13.

8. Ibid. 1:112.

9. Ibid.

10. Bethell, "Development of the Torpedo," 1:442.

11. *Dictionary of American Biography* 4:637.

12. U.S. Patent No. 36,654, 14 October 1862.

13. Hutcheon, *Robert Fulton*, 125–26; emphasis added.

14. U.S. Patent No. 35,285, 13 May 1862.

15. U.S. Patent No. 35,957, 22 July 1862.

16. U.S. Patent No. 41,112, 5 January 1864.

17. U.S. Patent No. 49,300, 8 August 1865.

18. Gray, *Devil's Device*, re: German failures, 222–24; re: American failures, 229–35.

19. U.S. Patent No. 58,661, 9 October 1866.

20. U.S. Patent No. 156,320, 27 October 1874.

21. Bethell, "Development of the Torpedo," 1:442.

22. Gray, *Devil's Device*, 111–14. *Ironclads in Action* also includes detailed accounts of the Russo-Turkish War.

23. Bethell, "Development of the Torpedo," 1:443. Bethell and H. W. Wilson (*Ironclads in Action*) do not always agree over details. News reports in the *Times* tend to be both inaccurate and partisan. When details differ, I have preferred to follow Bethell's version, although he does not always give his sources.

24. Gray, *Devil's Device*, 111–12.

25. *Times*, 9 July 1887.

26. Woods, *Spunyarn*, 42–44.

27. The story of the Secret will be found in Gray, *Devil's Device*, 50 and 57–58.

28. *Times*, 30 January 1878.

29. I possesses a photocopy of the Whitehead factory's consignment ledger, which ran from 26 January 1875 until 1915, when the entire works were evacuated to St. Pöltoen, forty miles south of Vienna, to escape bombing raids.

30. *Times*, 31 January 1878.

228 · *Notes to Pages 124–37*

31. Gray, *Devil's Device*, 114.

32. *Times*, 31 January 1875.

33. U.S. Patent No. 199,841, 29 January 1878.

34. U.S. Patent No. 223,855, 27 January 1880.

35. U.S. Patent No. 249,192, 8 November 1881.

36. U.S. Patent No. 202,453, 16 April 1878.

37. U.S. Patent No. 399,516, 12 March 1889.

38. Bethell, "Development of the Torpedo," 1:443. The spelling of Chinese names and places varies considerably from source to source and is often determined by the date of the publication in which they appear. So far as possible I have tried to follow more recent spellings.

39. Copies of many official reports concerning Civil War spar torpedo attacks can be found in chapter 19 of von Scheliha's *Treatise on Coast Defence*.

Chapter 7. A Perfect Nightmare

1. Roland, *Underwater Warfare*, 101.

2. Roland, *Underwater Warfare*, 102.

3. Ibid.; and Gray, *Devil's Device*, 67.

4. U.S. Naval Institute reproduced in Hutcheon, *Robert Fulton*.

5. Gray, *Devil's Device*, 68–73.

6. Ibid., 68–69.

7. Hutcheon, *Robert Fulton*, 48.

8. Ibid.

9. Ibid., 72–78.

10. Ibid., 84–86.

11. Ibid., 96–97.

12. Fulton's report to President Thomas Jefferson, dated 28 July 1807, quoted in Hutcheon, *Robert Fulton*, 96.

13. Fulton, *Torpedo War*.

14. Hutcheon, *Robert Fulton*, 124–25.

15. Bethell, "Development of the Torpedo," 1:443.

16. *Engineering*, 2 June 1871, 384.

17. Ibid., 1 October 1869.

18. Ibid., 18 March 1870.

19. PRO ADM 116/163.

20. *Engineering*, 18 March 1870; emphasis added.

21. Harvey, *Instructions*, 6 and 7.

22. Ibid., 10.

Notes to Pages 137–48 · 229

23. Ibid.

24. The date was March 1870 as confirmed in *Engineering,* 13 January 1871, 35.

25. Preston, *Battleships,* 16.

26. *Engineering,* 18 March 1870, n.p.

27. Gray, *Devil's Device.* The 1870 trials are the subject of chapter 1.

28. *Times,* 22 May 1872.

29. Marder, *Fear God and Dread Nought* (Fisher's letters), 1:73.

30. *Engineering,* 8 April 1870.

31. *Engineering,* 13 January 1871.

32. Sleeman, *Torpedoes and Torpedo Warfare,* 1880 ed., 130–31.

33. This periodical has not been traced but could be an internal service journal as the "broad arrow" was a government mark stamped onto tools, weapons, and other portable items. The mark remains in use today.

34. *Engineering,* 17 February 1871.

35. Ibid.; emphasis added.

36. Ibid., 28 July 1871, 63.

37. Harvey, *Instructions,* 5.

38. Ibid., 3 and 4.

39. *Engineering,* 13 January 1871, 35.

40. Vavasseur obituary in *Times,* 25 November 1908, p. 20.

41. Harvey, *Instructions,* 29.

42. *Engineering,* 1 December 1871, 360.

43. Mackay, *Fisher of Kilverstone,* 252.

44. *Times,* 16 November and 25 November 1908, obituaries.

45. Quoted by Mackay, *Fisher of Kilverstone,* 309.

46. Ibid. 344.

47. Ibid. 345.

48. *Times,* 25 March 1872, p. 5.

49. *Engineering,* 24 May 1872, 352, and 21 June 1872, 412.

50. *Engineering,* 24 May 1872, 352.

51. Harvey, *Instructions,* 27–28. Whether this was the same Captain McEvoy who invented the clockwork-powered 10-inch weapon referred to at the beginning of chapter 9 is not known, although it seems quite possible.

52. *Times,* 27 September 1873, p. 5.

53. Bethell, "Development of the Torpedo" 1:443.

54. Sleeman, "Torpedoes," 488.

55. *Illustrated London News Supplement,* 5 May 1877.

230 · Notes to Pages 150–54

56. *Times*, 7 August 1877, p. 9, and 14 August 1877, p. 8.

57. *Engineer*, 12 April 1889, 305.

Chapter 8. The Wizard of Oz

1. The origin of Schwartzkopff's pirated copy of the Whitehead torpedo related on pages 94–95 of Gray, *Devil's Device*, and based upon the anecdotal recollections of the late Count Balthazar Hoyos, is now thought to be incorrect. It is fairly certain that influential admirals in the Imperial German navy allowed Schwartzkopff to examine the initial batch of Whitehead torpedoes when they were delivered and encouraged him to copy the weapons in every detail. This theory has the support of Sleeman (*Torpedoes and Torpedo Warfare*, 493). It is possible that the burglary to which Hoyos referred in the interview with the author in 1972 took place much later and was concerned with the adoption of the gyroscope, although it is by no means certain that either Louis Schwartzkopff or the Berlin factory were involved in this particular example of nineteenth-century industrial espionage.

2. These details come from the Whitehead company's consignment ledger, a copy of which is held by the author.

3. Gray, *Devil's Device*, 130–33.

4. Extracted from a lecture by Robert Graham to the Royal Aeronautical Society quoted by Robert Millar in his doctoral thesis on Australian technology. I have used this to confirm facts and dates in this section.

5. Tomlinson, *Louis Brennan*. I corresponded with the late Norman Tomlinson during the preparation of the second edition of *Devil's Device*, and much of the information in this section came from him.

6. William Charles Kernot came from French stock and was born in the English county of Essex on 16 June 1845, he was only six years old when his parents emigrated to Australia in 1851, and he graduated from the University of Melbourne in 1864, gaining his master's degree in 1866. He also enjoyed the distinction of being the university's first qualified engineer.

Kernot joined government service in 1865, holding a variety of engineering appointments, latterly in the water supply office. In addition, he worked as a part-time lecturer at the university from 1868, when he began to specialize in civil engineering. By 1883 he had become Melbourne's professor of engineering, the first graduate of his alma mater to be appointed to a chair at the university.

His work with Louis Brennan commenced in 1876 and continued for many years. In this connection it should be noted that he visited Europe in 1878, 1891, and 1901. It seems probable he met with Brennan during the latter two trips and

Notes to Pages 155–71 · 231

that the two men discussed the technical problems which the inventor had encountered with his wire-propelled weapon.

Kernot served on a number of official boards of inquiry, including the New South Wales Royal Commission on Railway Bridges. A prominent and busy member of various professional societies and institutions, he nevertheless found time to become an inaugural member of the Australian Antarctic Committee. Working with two colleagues, he introduced electric lighting to the city of Melbourne, and his wide-ranging personal interests included velocipedes, steam cars, and hot-air ballooning. He died on 14 March 1909.

7. Kitson, "Brennan Torpedo," vol. 107, pt. 1, no. 1; vol. 107, pt. 2, no. 2; vol. [?], pt. 3, no. [?]. The patent number is given as 2105, 4 September 1875.

8. British Patent No. 3359. Letters patent sealed 1 February 1878 but backdated to 4 September 1877.

9. Kitson, "Brennan Torpedo," pt. 1, 74.

10. Ibid., 75. Millar gives length as sixteen feet.

11. See chapter 4.

12. Kitson, "Brennan Torpedo," pt. 1, 77.

13. Millar, *International Influence,* 158.

14. Ibid.

15. Kitson, "Brennan Torpedo," pt. 1, 77.

16. Ibid., 159.

17. Rennick, *Papers and Discussions* 3:70–73.

18. Gray, *Devil's Device,* 37.

19. Baker-Brown, *History of Submarine Mining,* 262; quoted by Millar, *International Influence,* 158.

20. Millar, *International Influence,* 158; emphasis added.

21. Probably the most comprehensive account of the Brennan forts, including site plans and photographs of the stations as they are today, will be found in *The Brennan Torpedo* by Alec Beanse, a booklet published by the Palmerston Forts Society in 1997. Further extensive details are recorded by Kitson in part 3, and these also include site plans and drawings.

22. Tomlinson, *Louis Brennan.*

23. *Times,* 6 September 1869.

24. Gray, *Devil's Device,* 156.

25. U.S. Patent No. 311,325, 27 January 1885.

26. Career details extracted from various obituaries and confirmed from the *Dictionary of American Biography,* vol. 5, pt. 1, pp. 303–4.

27. Sleeman, *Torpedoes and Torpedo Warfare,* 493.

28. *Engineer,* 17 August 1888, 144–45.

29. Ibid.

30. Carr, *Naval Pocket Book.*

31. A detailed technical account of the U.S. Navy's service torpedoes will be found in *Proceedings,* 19, no. 1, 1893.

Chapter 9. This Risk of Premature Explosion

1. Millar, *International Influence,* 185.

2. Beanse, *Brennan Torpedo,* 4 and 5.

3. U.S. Patent No. 327,380, 29 September 1885.

4. *Torpedo Manual,* 1929, 11.

5. Ibid.

6. U.S. Patent No. 134,493, 31 December 1872; emphasis added.

7. U.S. Patent No. 364,364, 7 June 7 1887.

8. U.S. Patent No. 245,864, 16 August 1881.

9. U.S. Patent No. 358,471, 1 March 1887.

10. U.S. Patent No. 370,570, 27 September 1887.

11. U.S. Patent No. 359,313, 15 March 1887.

12. See chapter 3.

13. British Patent No. 16,693, 1888. Also No. 4494, 20 September 1883; U.S. Patent No. 321,579, 17 February 1885.

14. Sleeman, "Torpedoes," 496.

15. Kirby, "History of the Torpedo," vol. 27, pt. 1, p. 38.

16. Sleeman, "Torpedoes," 496.

17. Kirby, "History of the Torpedo," vol. 27, pt. 1, p. 42.

18. See chapter 5.

19. Gray, *Devil's Device,* 220–21.

20. Bethell, "Development of the Torpedo," 8:244.

21. Sleeman, "Torpedoes," 496.

22. U.S. Patent No. 319,633 9, June 1885; U.S. Patent No. 375,417, 21 April 1891; British Patent No. 2614, 25 May 1883.

23. Sleeman, "Torpedoes," 496–97.

24. *Engineer,* 29 January 1892, 91.

25. Ibid.

26. *Engineer,* 12 February 1892, 123.

27. Kirby, "History of the Torpedo," vol. 27, pt. 1, p. 38.

28. *Engineer,* 19 February 1892, 150.

29. Ibid.

Notes to Pages 192–207 · 233

30. Ibid.

31. *Engineer*, 15 April 1892.

32. Ibid.

Chapter 10. An Eminently Humane Weapon

1. U.S. Patent No. 420,406, 28 January 1890.

2. British Patent No. 9774, 24 June 1891.

3. *Engineering*, 29 August 1890, 246.

4. *Engineer*, 29 August 1890, 172.

5. Australian Archives (Victoria), Commonwealth Record Series B3756, Victoria Department of Defence, Item No. 1891/3662 (hereafter cited as Australian Archives, CRS).

6. U.S. Patent No. 442,327, 9 December 1890.

7. *Engineer*, 29 August 1890, 246.

8. U.S. Patent 442,327, 9 December 1890.

9. *Engineer*, 9 September 1892, n.p.

10. British Patent No. 10581, 13 September 1890.

11. U.S. Patent No. 431,210, 1 July 1890.

12. Australian Archives, CRS, 1892/384.

13. Ibid.

14. Australian Archives, CRS, 1891/3662.

15. Ibid.

16. John Philip Holland is generally acknowledged as the inventor of the modern submarine. See Gray, *Few Survived*, 35 and 40–47.

17. Bethell, "Development of the Torpedo," 7:244.

18. U.S. Patent No. 453,861, 9 June 1891.

19. U.S. Patent No. 627,312, 20 June 1899.

20. *Age*, 26 September 1895, p. 5.

21. Millar, "International Influence," 185.

22. *Age*, 26 September 1895, p. 5.

23. U.S. Patent No. 638,463, 5 December 1899.

24. *Age*, 26 September 1895, p. 5.

25. U.S. Patent No. 559,711, 5 May 1896.

26. U.S. Patent No. 227,637, 18 May 1880.

27. U.S. Patent No. 632,089, 29 August 1899.

28. U.S. Patent No. 641,787, 23 January 1900.

29. Bethell, "Development of the Torpedo," 5:530, 28 December 1945.

30. U.S. Patent No. 1,683,085, 23 February 1921.

234 · *Notes to Pages 207–12*

31. Bethell, "Development of the Torpedo," 5:530.

32. Eliphalet Williams Bliss (1836–1903); see entry in *Dictionary of American Biography*, 1929 ed., 2:371.

33. See chapter 8.

34. Photocopy of ledger in author's collection.

35. Bethell, "Development of the Torpedo," 3:303.

36. No details. Copy in my collection lacks cover and title page but is dated 1914.

37. Milford, "U.S. Navy Torpedoes."

38. Letter from Robert Whitehead to his brother William, 19 July 1891, photocopy in author's collection.

39. This could be a reference to Guglielmo Marconi (1874–1937), who proposed a wireless-controlled torpedo to the British secretary of state for war, the Marquess of Lansdowne, in a letter dated 20 May 1896 (PRO W032/8594). The Italian inventor claimed to have "discovered electrical devices which enable me to guide or steer a self-propelled boat or torpedo from the shore, or from a vessel, without any person being on board the said boat or torpedo." In brief, the remote control apparatus operated by driving a "pecker" motor one step at a time whenever the transmitting Morse key was depressed. Marconi claimed that his device would work even when enclosed in a box under water, a claim O'Dell considered to be wholly false. Baker similarly took the view that the assertion was completely "beyond the capabilities of the Marconi apparatus at that time." Marconi was to take out the world's first patent for wireless telegraphy on 2 June 1896 (No. 12039), less than a fortnight after his letter to Lansdowne. And he successfully demonstrated the transmission of signals without the use of wires on Salisbury Plain later the same year. However, nothing more was heard of the proposed radio-controlled torpedo. For further details and a technical description of the 1896 patent, see T. H. O'Dell, *Inventions and Official Secrecy*, 50–57, and W. J. Baker, *A History of the Marconi Company*.

It has been claimed that Nikola Tesla (1856–1943) overcame the failings of Marconi's torpedo in his U.S. Patent No. 613809, dated 8 November 1898, which incorporated an aerial to receive radio signals, which enabled the device to be steered remotely without the use of wires or cables.

Such claims were, however, misleading, for Tesla himself described his apparatus as a "boat or . . . floating vessel" and it is clear from both the patent's text and its supporting drawings that it was only a surface weapon and not a torpedo as understood by the term in the closing years of the nineteenth century. It was no more a torpedo than a Civil War *David* was a submarine. Indeed, the analogy is aptly close, for neither contrivance could dive or be controlled under water.

Tesla was born in Croatia and came to the United States in 1884. He was, like Marconi, an important inventor in the field of electrical engineering and was the first scientist to find an effective way to utilize alternating current.

40. *Engineer*, 25 October 1901.

41. Ibid.

Bibliography

The research for this book began while I was preparing the original edition of *The Devil's Device* and has been continuous, if spasmodic, since 1972. Interviews that took place as long as thirty years ago (and the contemporary notes) have been held on file as part of my research library of torpedo development since that time.

Public Documents

Australian Archives (Victoria) Commonwealth Record series. [Victoria, Department of Defence Section] 91/1892, 269/1892, 384/1892.

British Library, London. Patents.

Public Record Office, London. ADM 1/580, ADM 116/135, ADM 116/146, ADM 116/163, ADM 116/164, WO 32/8594.

Reports, Monographs, and Lectures

Admiralty Torpedo Committee. *Final Report.* July 1876. A preliminary report had been issued in May 1873. PRO ADM 116/163.

Alliman, A. L. Lectures given at King's College Engineering Society in 1877 and January 1879. No other details known.

Baker-Brown, Lt. Col. W. *History of Submarine Mining in the British Army.* Chatham, Kent: Royal Engineers, 1910.

Barber, Lt. Francis Morgan, USN. Lecture on the Whitehead Torpedo. 20 November 1874. U.S. Navy Torpedo Station, Newport, R.I.

Berdan Torpedo. General Information Series No 6. Office of Naval Intelligence, U.S. Navy, June 1887.

Bradford, Lt. Cdr. Royal B., USN. *Notes on Movable Torpedoes.* Newport, R.I.: U.S. Navy Torpedo Station, 1882.

Drake, Lt. Franklyn J., USN. Three lectures delivered to U.S. Naval War College, 26–28 October 1892. Reprinted in the *Proceedings of the United States Naval Institute* 19, no. 1 (1893).

238 · Bibliography

Fulton, Robert. *Torpedo War and Submarine Explosions.* New York: n.p., 1810.

Harvey, Cdr. Frederick. *Instructions for the Management of the Harvey Sea Torpedo.* London: E. & F. N. Spon, 1871.

Kernot, William C. Commencement address to Melbourne University students. 1893. Original source unknown.

Rennick, W. R., ed. *Papers and Discussions of the Victorian Institute of Engineers.* Vol. 3. Melbourne, c. 1903.

Royal Navy Torpedo Manual. London: HMSO, 1887.

Royal Navy Torpedo Manual. London: HMSO, 1929.

Scheliha, Lt. Col. [Victor] von. *A Treatise on Coast Defence.* London: E. & F. N. Spon, 1868.

Southern Historical Society Papers 4, no. 5 (November 1877): n.p.

Wilson, Robert. *The Screw Propeller–Who Invented it?* 2d ed. Glasgow: Thomas Murray & Son, 1880.

Unpublished Material

Kirby, Geoff. "The Development of Rocket-propelled Torpedoes." N.d.

Millar, Robert. "The International Influence of Australian Technology 1850–1910." Doctoral thesis. N.d.

Whitehead and Company, Fabbrica di Torpedini. Consignment Ledger covering period 26 January 1875 to 1916 and serial numbers 1 to 14019. Photocopy.

Books, Periodicals, and Newspapers

Armstrong, G. E. *Torpedoes and Torpedo Vessels.* London: Bell, 1901.

Bacon, Adm. Sir Reginald. *Lord Fisher.* 2 vols. London: Hodder & Stoughton, 1929.

Beanse, Alec. *The Brennan Torpedo.* Fareham, Hampshire: Palmerston Forts Society, 1997.

Bethell, Cdr. Peter. "The Development of the Torpedo." A series of articles published in *Engineering.* Pts. 159–61, 25 May 1945–15 March 1946.

Boase, Frederick. *Modern English Biography.* Vol. 5, D–K. London: Frank Cass, 1965.

Bourne, John. *A Treatise on the Screw Propeller.* N.p.: N.p., 1852.

Caseli, Antonio, and Marina Cattaruzza. *Sotto i mari del mondo–La Whitehead 1875–1990.* Italy: Editori Laterza, 1990.

Church, William Conant. *Life of John Ericsson.* 2 vols. London: Sampson Low, Marston, Searle & Rivington, 1890.

Clowes, Sir W. Laird. *Four Modern Naval Campaigns.* London: Hutchinson, 1906.

Fisher, Adm. Lord John. *Records*. London: Hodder & Stoughton, 1919.

Gray, Edwyn. *The Devil's Device*. 2d ed. Annapolis: Naval Institute Press, 1991.

———. *Few Survived*. 2d ed. London: Leo Cooper; New York: Barnes & Noble, 1996.

Hough, Richard. *First Sea Lord*. London: Allen & Unwin, 1969.

Hutcheon, Wallace, Jr. *Robert Fulton: Pioneer of Undersea Warfare*. Annapolis: Naval Institute Press, 1981.

Kemp, Lt. Cdr. P. K., ed. *The Fisher Papers*. 2 vols. London: Naval Records Society, 1960.

Kirby, G. J. "A History of the Torpedo." 4 pts. *Journal of the Royal Naval Scientific Society* 27, nos. 1 and 2. N.d.

Kitson, Michael. "The Brennan Torpedo." *Royal Engineers Journal* 107, pt. 1, no. 1; pt. 2; pt. 3 (vol. and no. unknown), n.d.

Mackay, Ruddock R. *Fisher of Kilverstone*. Oxford: Oxford University Press, 1973.

Marcot, Roy M. *Civil War Chief of the Sharpshooters Hiram Berdan, Military Commander and Firearms Inventor*. Irvine, Calif.: Northwood Heritage Press, 1989.

Marder, Arthur J., ed. *Fear God and Dread Nought*. Vol 1. London: Jonothan Cape, 1952.

Milford, Frederick J. "U.S. Navy Torpedoes." In *The Submarine Review*. Annandale: Naval Submarine League, c. 1996.

O'Dell, T. H. *Inventions and Official Secrecy*. Oxford: Clarendon Press, 1994.

Poland, Rear Adm. E. N. *The Torpedomen*. Emsworth: Kenneth Mason Publications, c. 1986.

Preston, Antony. *Battleships, 1856–1977*. London: Phoebus Publishing, 1977.

Roland, Alex. *Underwater Warfare in the Age of Sail*. Bloomington: Indiana University Press, 1978.

Scharf, Thomas J. *History of the Confederate States Navy*. Facsimile ed. New York: Fairfax Press, 1977.

Sleeman, Lt. Cdr. C. W. "Torpedoes." In *The Naval Annual*. 1887 ed. Portsmouth: Griffin, 1888.

———. *Torpedoes and Torpedo Warfare*. 1st and 2d ed. Portsmouth: Griffin, 1880, 1889.

Sueter, Cdr. Murray F. *The Evolution of the Submarine Boat, Mine, and Torpedo*. Portsmouth: Griffin, 1907.

Tomlinson, Norman. *Louis Brennan*. Chatham, Kent: John Howell Publications, 1980.

240 · *Bibliography*

Wellesley, Col. the Hon. Frederick Arthur. *Recollections of a Soldier-Diplomat.* Edited by Sir Victor Wellesley. London: Hutchinson, c. 1931.

Wilson, H. W. *Ironclads in Action.* 2 vols. 4th ed. London: Sampson Low, Marston, 1896.

Woods, Vice Adm. Sir Henry F. *Spunyarn.* N.d.

Annual Publications and Standard Reference Works

Australian Dictionary of Biography. Edited by B. Nairn, G. Serle, and R. Ward. Vol. 5, 1851–90. Melbourne: Melbourne University Press, 1974.

Burkes Peerage, Barontage, and Knightage. 1928 ed.

Dictionary of American Biography. 1930 ed. Edited by Dumas Malone. New York: Charles Scribner's Sons.

Dictionary of National Biography. 1921 ed. Oxford: Oxford University Press.

Naval Pocketbook. London: W. Thacker, 1899.

Who Was Who. Vol 3. N.p: n.d., 1929–40.

Index

Abbot, General, 187

Abtao, 52

Adams, 169

Admiralty Torpedo Committee, 26–28, 43–44, 47, 135, 147

Albermarle, 42, 109–13

Alexander, Andrew, 16

Alexandrovsky, Ivan F., 37–40

Alliman lectures, 26, 85

Almirante Cochrane, 53

Alsbau, Mary Josephine, 203

Amethyst, 53

Andrada, 186

Andrew Wodehouse, 142

Assar-i-Cheuket, 119

Atalanta, 170

Balfour, Arthur James, 146

Barber, Francis Morgan, 20–21

Barber, Francis Robert, 21

Barham, 212

Barlow, Joel, 114

Barrow, Thomas E., 205

battle tactics: Greek and Roman, 1; Middle Ages, 2; fourteenth to mid-seventeenth centuries, 2–3; mid-nineteenth century, 6; fireships, 6

Beauregard, Pierre G. T., 107, 113

Bennet, Henry Morden, 182–84

Berdan, Hiram: Borden, misspelled as, 85–86; career and death, 96–98, 102;

ironclad destroyer, 101; submarine gun, 100–101; torpedo specifications and tactical employment, 96, 98–99, 101; torpedo fails, 100–101; Turkey contract and trials, 98, 99–100; U.S. Navy, rejection by, 99

Blanco Encalada, 53

Bliss, Eliphat Williams, 208. *See also* Bliss and Williams

Bliss and Co. Fiume 45cm Mk 1 torpedo (under license), 209–10

Bliss and Williams (E. W. Bliss and Co.), 175, 207–10

Borden torpedo. *See* Berdan, Hiram

Bourne, John, 66

Braham, Phillip, 15–16, 92, 101

Brennan, Louis: Brennan Torpedo Company, 157–58; and British Army, 155, 156–47, 158; British government contract (1883), 159; British government contract (1887), 166–67; and Calvert, William, 157; death of, 168; early life, 152; education, 153; grant from Victoria government, 157; honors, 167; inventions, 153, 154, 168; invited to England, 157–58; and Kernot, 154–55, 161–66; Royal Engineers Committee reports, 158–59; and Royal Navy, 156–58; Sims-Edison torpedo, 191; and Smith, Alexander, 153, 156–57; and Temper-

[241]

Brennan, Louis (*continued*)
ley, 157–58, 161; and wire-propulsion principle, 154

Brennan torpedo: depth-keeping mechanism, 159, 164–66; forts built for, 153, 167–68; half-scale model (1876), 155; Hobson Bay demonstration (1879), 157; patent (1875), 155; patent (1877), 155–56; 1877 patent improved, 156, 166; press criticizes, 188–89; propellers, twin contra-rotating, adopted, 156; propulsion and guidance system, 155; Royal Engineers Museum, examples at, 155; sealed units (black boxes), 159, 165; sealed units' secrets revealed by Kernot, 161–66; security surrounding, 159–61, 165; Williamstown dock test, 156

Brotherhood air engine, 29

Bushnell, David, 5, 83–84, 129

Callender, Miles L., 21, 44, 101, 205

Calvert, William, 157

Camel, 138

Canseco, Diez, 52

Cave torpedo, 85

Cavett, William K., 194

Chambers, Washington Irving, 176

Chen-Yeun, 128

Chien-Chiang, 128

Childers, Hugh C. E., 158

Chile-Peru War (1879–82), 51–53

Christopher of the Tower, 2

Condell, Carlos, 52

Confederate Torpedo Bureau, 107

Connaught, Duke of, 189–90

Courbet, Rear Admiral, 127

Covadonga, 52

Cunningham, Patrick, 199–200

Curtis turbine engine, 211

Cushing, William Barker, 109–13

David-class torpedo boat, 108–9

Davidson, James, 80, 81–82

Decatur, Stephen, 133

de Ciotta, Giovanni, 7, 12

Delamater, C. H., 92, 114

de Luppis, Giovanni: and *Der Küstenbrander,* 7–8, 12; disappointment and death, 9; and Whitehead, 7–9

de Pierola, Nicholas, 55

Der Küstenbrander, 7–8, 12

di Persano, Carlo, 2

Donegal, 140

Dorothea, 132

Drudge, 189–90, 192

Edinburgh, Duke of, 146

Edison, Thomas Alva, 186, 192, 207. *See also* Sims-Edison torpedo

Ellery, Major, 195

Elliot, William H., 114–15

Ericsson, John: awash boat, 44; Bureau of Naval Ordnance, dispute with, 92; *Columbiad,* 114; and Davidson, 80–82; *Destroyer,* 91–94; early life, 63–64; emigrates to the United States, 77; in England, 64; and *Francis B. Ogden,* 76; illness and death, 68; and Lay's torpedo, 43–46; and *Monitor,* 67–68; pneumatic torpedo (tubular cable torpedo), 18, 69, 90–91; projectile torpedo (hydrostatic javelin), 89, 91–94; propeller, screw, 65, 66, 68–69, 76; propellers, contra-rotating, 20, 62–63, 66, 77; and *Robert F. Stockton,* 77; and Trimby, 67–68

Esmeralda, 52

Excellent, 13, 138, 140, 141

fireships, 6

Fisher, John (Jacky): Admiralty Torpedo Committee, advises, 27; and Harvey torpedo, 140, 145; Naval Ordnance, director of, 145; and private industry, 146; Royal Navy torpedo school, head of, 140; and Sims-Edison torpedo, 189; and Vavasseur, 145–46; and von Scheliha torpedo, 63; and Whitehead torpedo, 14–15, 140, 145

Fiume torpedoes: 14-inch Model A, 29, 80; 15-inch, 122–24; 16-inch standard, 8–9, 19, 29, 33, 45, 67; 18-inch, 9, 210; 21-inch, 195, 210; 45cm Mk 1 torpedo (under license), 209–10

Franco-China War (1884–85), 127–28

Fraser, Colonel G., 82

Fulton, Robert: and Boulogne, attack on, 131–32, 137; changes allegiance, 131; and *Columbiad* (torpedo gun), 114, 133; drifting mines, 132; early life and death of, 133, 134; grappled mines, 132; harpoon gun, 132–33; limpet mines, 5, 129–30, 132; *Nautilus*, 83, 129–31; riverine steam navigation, 134; screw propeller, 83; torpedo, self-propelled, 114; torpedo, spar, 106, 113–14, 133; torpedo, towed (submarine bomb), 129–31, 133

Fu Sing, 127

Gatling, Richard Jordan, 126–27

Gern, Ottomor B., 40

Gianibelli, Frederico, 4

Giese, William, 88–89

Glassell, William T., 107–9

Goschen, George Joachim, 81

Graham, Robert, 154, 168

Grau, Don Miguel, 52

Griffiths telescopic torpedo, 149

Grinder, 146

G-7e torpedo (German), 185

Haight, George E. 57. *See also* torpedoes

Hall steam-propelled torpedo, 95

Halpine, Nicholas, 200–201

Harvey, Frederick: agitates for financial recognition, 141, 147; John Harvey, confused with, 131, 141; promotion dispute, 141–42; and Yarmouth demonstration trials, 142

Harvey, John, 134

Harvey towed torpedo: Admiralty Torpedo Committee examines, 135–37; demonstrated at Spithead in 1870, 137–39; demonstrated at Spithead in 1872, 146; demonstrated at Yarmouth, July 1871, 142–43; demonstrated at Yarmouth, November 1871, 144; described, 135–37, 142–43; explosives used by, 135–37; Fisher supports, 139–40; and London Ordnance Company, 141–42; and Nunn, 135; press supports, 139, 147; and Royal Laboratory, 146, 148; and Royal Navy, 140–41, 148; Russia purchases, 135; tactical use, 137

Hassenpflug, Louis, 14

heaters: Armstrong's Elswick, 32; Bliss-Leavitt, 211; Hardcastle, 32; Lay, 45; Maxim, 206; Reynolds, 178; Royal Gun Factory, 206, 211; von Scheliha, 31–32; Whitehead, 32, 210, 211; Winsor, Wood, and Haight, 58

Heenan and Froud torpedo engine, 59, 196

Hertz, Heinrich, 212

Hewlett, R. S., 13

Horsey, Air Algernon de, 53

Housatonic, 109

Houston, William, 72–73

Howell, John Adams, 168–170

Howell flywheel torpedo: advantages over Whitehead weapon, 24, 173–74; azimuth and pendulum control system, 24; first design (1870) 21–25; first patent (1871), 169; flywheel propulsion, principles of, 23; flywheel weight and gyration speed, 171; Hotchkiss Ordnance Company of America, 170; model 1885 patent improved, 170–72; press criticizes, 188; U.S. Navy issue, 174; U.S. Navy Mark 1 torpedo, confused with, 22, 170; U.S. Navy torpedo Mod 94 performance, 172–73; U.S. Navy torpedo Mod 95 performance, 173

Hoyos, Georg, 123–24, 208

Huascar, 46, 52–54

Hunley, 109

Hunter, James, 72

Idajalieh, 119, 121

Independencia, 52

Ingraham, Commodore, 108

Intikbah, 124

Iowa, 174

ironclads, 3

Jaque's subaquatic torpedo, 196–97

Johnson, Alfred, 197–98

Johnson, Slacke, and Lacey torpedo, 204

Jeffers, William, 90–91, 92, 93

Jervois, William, 155

Just, Thomas Wemyss, 201–3

Kernot, William: assists Brennan, 154–55, 161–65; biographical details, 154n6; "black boxes" secret revealed by, 161–66; and Brennan Torpedo Company, 155

Ketcham, Isaac A., 113–14

Kirkland, William, 44–46

Knapp, Henry F., 58–59

Kniaz Constantine, 120–22

Krabb, H. K., 38

Laplace, Pierre-Simon de, 129

Lassoe, Valdemar, 196–97

Lauderdale, Earl of, 72

Lay, John: career, 41–43; death, 61; patents, 46; and Peruvian government, 43; and von Scheliha, 50. *See also* Lay-Haight torpedo; Lay torpedo

Lay-Haight torpedo, 46–47, 56, 58

Lay torpedo: development, 46; demonstrated at Antwerp, 51; demonstrated at Bosphorus, 54–55; demonstrated at Brightlingsea, 56, 59–61; demonstrated at Faversham, 55; Ericsson criticizes, 43–45; European sales campaign, 49; fails during Chile-Peru War, 46, 52–53; heater system, 45, 49–50; Lay-Haight torpedo, 46–47, 56, 58; moveable torpedo submarine, 43–46, 46–61; and Nordenfelt, 55; patent drawings, misleading, 49; patent (1879), 48–51; press criticizes, 61; propellers in bow mounting, 56–58; Russian government contract, 48, 61; U.S. Navy sales, 55; spar torpedo with Wood, 41–42, 110, 111

Leavitt, Frank McDowell: early career and patents, 207; Fiume samples, 208; turbine engines and weapon development, 211; U.S. Navy classification of Bliss-Leavitt torpedo, 209–10; and Whitehead, 208; work during World War I, and death, 211–12. *See also* torpedoes

Lee, Francis D., 107, 113

Lege torpedo, 150–51

Lodge, Oliver, 212

Lowe, Royal Laboratory foreman, 19,
81–82, 104

Mahmoudieh, 122
Maitland, Anthony, 72, 73
Makarov, Stepan, 121–22, 124
Mallory, William H., 124–25
Marconi, Guglielmo, 212n39
Maxim, Hiram, 175
Maxim, Hudson, 175–76, 205–7
McEvoy, C. A., 147
McEvoy, Captain, 175
McLean, James H., 88–89
McMullen, George W., 204–5
Menzing, Captain, 119, 149
Merriam, Scovil S., 116–17
Millar, Charles and Edwin, 157
Miller, Alfred P. S., 178
Milne, David, 73
Milton, Lord, 87–88
mines: in American Civil War, 6, 107;
Colt, 6; Fulton, 129–30, 132; Herz
horn, 6
Monarch, 131
Monde, Gaspard, 129
Monitor, 67–68
Mortenson, Hans, 126
Motorite explosive fuel, 206–7
Murphy, George Read, 194–96, 198

Nasmyth, James, 77
Nealey, Sid Hugh, 179–81
New Ironsides, 108–9
Nickolaevitch, Constantine, 38
Nordenfelt, Thorsten, 184–86
Nunn, William, 135, 144

Ocean, 140
Orling torpedo, 212

Pallas, 70

Pasha, Hobart, 54, 59, 100, 124
Patrick dirigible, 176–77
Paulson steam-propelled torpedo, 85,
94–95
Peck, Edward C., 94
Popov, Vice Admiral, 38
Porter, David Dixon, 101–2
Powhatan, 108
propellers, twin contra-rotating, 20,
62–63, 158. *See also* Ericsson, John;
Wilson, Robert
propulsion. *See* torpedoes
Puschschine, Lieutenant, 121

Quick, George, 16–17

Radford, Rear Admiral, 169
Rains, G. J., 107
ram, 1–2
Ramillies, 106, 114
Ramsay, George M., 115
Ramus, Charles Meade, 86–87
Rennick, W. R., 166
Ressell, Joseph, 65–66
Reynolds, George H., 178–79
Richmond, Duke of, 74
Rohestvensky, Zinovi, 121–22
Royal Laboratory experimental torpedo,
104
Royal Sovereign, 138–39, 146
Russo-Turkish War (1877–78), 118–24

Schwartzkopff, Louis, 152nl
screw propeller: advantages over paddlewheel, 67, 73; *Alecto* and *Rattler*
contest, 76; and Delisle, 66; origins
of, 66, 83–84; and Ressell, 65–66. *See
also* Ericsson, John; Smith, Francis
Pettit; Wilson, Robert
Seif, 119
Shah, 53

Shapovsky, Colonel, 40
Sims-Edison torpedo, 186–93
Singer, Morgan, 27
Sino-Japanese War (1894–95), 128
Sinope, 122, 124
Sleeman, C. W., 62, 82, 186, 188
Smith, Francis Pettit, 65, 66, 69, 75–76, 77, 84
Smith, Henry Julius, 178
Smith, Oliver C., 115–16, 125
Spanish-American War (1898), 170
Sullivan and Etheridge torpedo, 182–83

Tchesma, 122
Tegethoff, Wilhelm, 2
Temperley, John, 157–58, 161, 166–67
Tesla, Nikola, 212n40
Timby, Theodore R., 67–68
Ting, Admiral, 128
Tin Yuen, 128
Tirpitz, Alfred von, 152
Tombs, James H., 109
torpedo, 5, 107
torpedo boats, 127
torpedoes (by propulsion system):
　carbonic acid gas liquified: Knapp,
　　Henry F., 58–59; Lay-Haight, 46–
　　47, 58; Wood-Haight, 57–58
　carbonic and other gases: Patrick diri-
　　gible, 176–77. *See also* Lay, John
　*clockwork/springs: Der Küsten-
　　brander*, 7–8, 12; Mallory, William
　　H., 125; McEvoy, Captain, 175; Nea-
　　ley, Sid Hugh (rotary), 179–81
　compressed air: Alexandrovsky, Ivan
　　F., 37–40; Barber, Francis Morgan,
　　20–21; Bliss and Co. Fiume 45cm
　　Mk 1 (under license), 209–10;
　　Gern, Ottomor B., 40; Halpine,
　　Nicholas, 201; Milton, Lord, 87–88;
　　Murphy, George Read (*Victoria*),

194–96, 198; Reynolds, George H.,
178–79; Schwartzkopff, Louis,
152nl; Smith, Henry Julius, 178;
U.S. Navy Torpedo Station 14-inch,
20; van Bibber, Andrew, 201; Warsop and Brentall, 12–13; Woolwich
FL 16-inch Mk 1, 19, 27, 82, 83;
Woolwich RGF 21-inch, 195; Woolwich RL 14-inch Mk VIII, 59
　electrical: Barrow, Thomas E., 205;
　　German G-7e, 185; Halpine storage
　　battery, 200–201; Johnson, Alfred,
　　197–98; Nordenfelt, Thorsten,
　　184–86; Sims-Edison, 186–93; Willi-
　　ams, J. S., 104–5, 185
　explosive ram: Cavett, William K., 194;
　　Uhlan, 117–18
　flywheel. See Howell, John Adams
　jet-propelled, 95
　magnetic: Johnson, Slacke, and Lacey,
　　204; McMullen, George W., 204–5
　projectile: Braham, Phillip, 15–16, 92,
　　101; Fulton *Columbiad*, 114, 133;
　　Fulton harpoon gun, 132–33;
　　Jaque's subaquatic, 196–97; Just,
　　Thomas Wemyss, 201–3; Lassoe,
　　Valdemar, 196–97. *See also* Ericsson,
　　John
　rocket and rocket turbine: Alexander,
　　Andrew, 16; Barber, Francis Robert,
　　21; Callender, Miles L., 21, 44, 101,
　　205; Chambers, Washington Irving,
　　176; Cunningham, Patrick, 199–200;
　　Giese, William, 88–89; Mallory, Wil-
　　liam H., 124–25; Ramus, Charles
　　Meade, 86–87; Royal Laboratory ex-
　　perimental, 104; Shapovsky, Colo-
　　nel, 40; Sullivan and Etheridge,
　　182–83; Weeks, Asa, 98, 101, 103–4;
　　Willoughby, James D. 115–16. *See
　　also* Berdan, Hiram

spar: chemical fuses, 108; combat use, 106, 107–10, 119–22, 127–28; Elliot, William H., 114–15; Gatling, Richard Jordan, 126–27; Griffiths telescopic, 149; Ketcham, Isaac A., 113–14; Lee, Francis, 107, 113; Mallory, William H., 124–25; Merriam, Scovil S., 116–17; Mortenson, Hans, 126; origins of, 106–7, 113, 133; Ramsay, George M., 115; Smith, Oliver C., 115–16, 125; van Drebbel, Cornelius, 4–5; von Court, Charles J., 117; Weeks, Asa, 98, 101, 103–4; Wood-Lay, 41–42, 110, 111. *See also* Lay, John; Lay torpedo

steam: Hall, 95; Paulson, 85, 94–95; Peck, Edward C., 94

towed: American, French, and German, 148; Griffiths, 149; Lege, 150–51; Menzing, Captain, 119, 149. *See also* Harvey, Frederick; Harvey, John

wire: Maxim, Hudson, 175–76. *See also* Brennan, Louis

wireless (radio) controlled: Marconi, Guglielmo, 212n39; Orling, 212; Tesla, Nikola, 212n40

unidentified systems: Bennet, Henry Morden, 182–84; Cave, 85; explosive propellant (Hudson Maxim), 205; McLean, James H., 88–89; Miller, Alfred P. S., 178; Quick, George, 16–17

Torpedoist, 119

Trenholm, George A., 108

turbine power units, 211

Turtle, 5, 83, 129

Tzarevitch, 119

Uhlan ram torpedo, 117–18

United States Navy and the Whitehead torpedo, 20, 169, 207, 208–10

United States Navy torpedo Mod 94, 172–73

United States Navy torpedo Mod 95, 173

United States Projectile Company, 208

van Bibber, Andrew, 201

van Drebbel, Cornelius, 4–5

Vavasseur, Josiah: and Harvey torpedo, 144–145; and hydraulic recoil cylinder, 145; and London Ordnance Company (Fisher), 145–46; and Sims-Edison torpedo, 188; and Sir W. G. Armstrong Whitworth and Co., 145

von Court, Charles J., 117

von Scheliha, Victor: bankruptcy and imprisonment, 37; Confederate Army service, 26, 63; and Ericsson torpedo, 30; and heater principle, 31–32, 49; and Lay, 29, 49; other patents, 30, 35; torpedo built in Russia, 29, 36; torpedo examined by Admiralty Torpedo Committee, 26, 32–33; torpedo provisional patent (1872), 28–32; torpedo patent (1873), 33–36; and Wellesley, Frederick, 26, 28–29, 36–37; Whitehead's vs. von Scheliha's weapon, 30–31

Vernon, 191, 212

Virginia, (ex-*Merrimac*), 68

Warsop and Brentall torpedo, 12–13

water petard, 5

Weeks, Asa, 98, 101, 103–4

Wellesley, Frederick Arthur: diplomatic career, 28; and von Scheliha, 26, 28–29; witnesses torpedo tests in Russia, 29

Williams, J. S., 104–5, 185

Willoughby, James D., 115–16

Whitehead, Robert: and Bliss and Company, 208–211; career similar to Ericsson's, 64–65; contract and tests with Austria, 2, 9; contract with British government (1871), 9, 140; contract with Russia (1876), 123; contract with Turkey and other negotiations, 123–24; and de Luppis, 7–9; England orders two hundred torpedoes, 80; fish torpedo, criticisms of, 13–15, 17–18; Fiume factory output, 9, 152; gyroscope, 102; inventor of modern torpedo, 8–10, 11; and Leavitt, 207–9; and propellers, contra-rotating, 27, 63, 80–81; "the Secret," 9, 164–66; torpedo, first prototype, 8–9, torpedo, Fiume 14-inch Model A, 29, 80; torpedo, Fiume 15-inch, 122–24; torpedo, Fiume 16-inch standard, 8–9, 19, 29, 33, 45, 67; torpedo, Fiume 18-inch, 9, 210; torpedo, Fiume 21-inch, 195, 210; turbine experiments, 211; United States interest (1869), 169; United States, special delivery of seven torpedoes (1913), 210; Woolwich-built torpedo, test of first, 81–82

Wilson, Robert: British government, monetary award from, 63, 77, 80; character, 75, 78, 79, 83; early working life, 70; and Ericsson, 78, 83; experimental work (1821–32), 71–74;

Leith trials, 73; miter wheels and bevel gears, 78, 80, 85; monograph, 66–67; and Nasmyth, 77; propeller, screw, 70; propeller, screw, dismissed by British Admiralty, 74; propeller, screw, not patented, 75; propeller, screw, publicly tested (1827), 72–73; propellers, twin contra-rotating, 1876 patent for, 78–80; propellers, twin contra-rotating, Royal Laboratory tests and adopts, 63, 82–83; Smith, Pettit, suspicions concerning, 74–75

Winsor, William E., 57

Wood, William H., 57

Wood, William W. W., 41

Wood-Haight torpedo, 57–58

Woods Pasha, Henry, 123

Woolwich Royal Laboratory, 19, 82, 204

Woolwich torpedoes: FL 16-inch Mk 1, 19, 27, 82, 83; RGF 21-inch, 195; RL 14-inch Mk VIII, 59

Xenia, 119

Yang Wei, 128
Yang Wu, 127
Yu Yen, 128

Zalinsky dynamite gun, 168
Zazarenni, Lieutenant, 121

ABOUT THE AUTHOR

Born in London in the days when ocean freighters unloaded their exotic cargoes in the heart of the capital, Edwyn Gray has had a lifelong interest in ships and the sea. He counts among his forebears John Philipot, who led a makeshift English fleet to victory in the fourteenth century; another of his ancestors fought at Trafalgar. Gray was educated at the 450-year-old Royal Grammar School, High Wycombe, and he went on to read economics at London University before joining the British civil service as a professional-grade officer.

Gray began writing magazine features in 1953, and his first book was published in 1969. He became a full-time author in 1980. All of his books have been published in both Britain and the United States, with a number translated into Dutch, Italian, German, Danish, Swedish, and Japanese. Respected for his various works on submarine warfare, his biography of Robert Whitehead, *The Devil's Device*, the second edition of which was published by the Naval Institute Press in 1991, established his reputation as the world's leading–and virtually only–torpedo historian.

Edwyn Gray now lives in Norfolk, England–the country in which Adm. Horatio Nelson was born–and divides his time between writing, research, and consultancy work for the international media.

The Naval Institute Press is the book-publishing arm of the U.S. Naval Institute, a private, nonprofit, membership society for sea service professionals and others who share an interest in naval and maritime affairs. Established in 1873 at the U.S. Naval Academy in Annapolis, Maryland, where its offices remain today, the Naval Institute has members worldwide.

Members of the Naval Institute support the education programs of the society and receive the influential monthly magazine *Proceedings* and discounts on fine nautical prints and on ship and aircraft photos. They also have access to the transcripts of the Institute's Oral History Program and get discounted admission to any of the Institute-sponsored seminars offered around the country.

The Naval Institute also publishes *Naval History* magazine. This colorful bimonthly is filled with entertaining and thought-provoking articles, first-person reminiscences, and dramatic art and photography. Members receive a discount on *Naval History* subscriptions.

The Naval Institute's book-publishing program, begun in 1898 with basic guides to naval practices, has broadened its scope to include books of more general interest. Now the Naval Institute Press publishes about one hundred titles each year, ranging from how-to books on boating and navigation to battle histories, biographies, ship and aircraft guides, and novels. Institute members receive significant discounts on the Press's more than eight hundred books in print.

Full-time students are eligible for special half-price membership rates. Life memberships are also available.

For a free catalog describing Naval Institute Press books currently available, and for further information about subscribing to *Naval History* magazine or about joining the U.S. Naval Institute, please write to:

Membership Department
U.S. Naval Institute
291 Wood Road
Annapolis, MD 21402-5034
Telephone: (800) 233-8764
Fax: (410) 269-7940
Web address: www.navalinstitute.org